Adopting Circular Economy Current Practices and Future Perspectives

Adopting Circular Economy Current Practices and Future Perspectives

Special Issue Editor

Idiano D'Adamo

MDPI • Basel • Beijing • Wuhan • Barcelona • Belgrade • Manchester • Tokyo • Cluj • Tianjin

Special Issue Editor
Idiano D'Adamo
Unitelma Sapienza—University
of Rome
Italy

Editorial Office
MDPI
St. Alban-Anlage 66
4052 Basel, Switzerland

This is a reprint of articles from the Special Issue published online in the open access journal *Social Sciences* (ISSN 2076-0760) (available at: https://www.mdpi.com/journal/socsci/special_issues/Adopting_Circular_Economy).

For citation purposes, cite each article independently as indicated on the article page online and as indicated below:

LastName, A.A.; LastName, B.B.; LastName, C.C. Article Title. *Journal Name* **Year**, *Article Number*, Page Range.

ISBN 978-3-03928-342-2 (Pbk)
ISBN 978-3-03928-343-9 (PDF)

Cover image courtesy of Idiano D'Adamo.

© 2020 by the authors. Articles in this book are Open Access and distributed under the Creative Commons Attribution (CC BY) license, which allows users to download, copy and build upon published articles, as long as the author and publisher are properly credited, which ensures maximum dissemination and a wider impact of our publications.

The book as a whole is distributed by MDPI under the terms and conditions of the Creative Commons license CC BY-NC-ND.

Contents

About the Special Issue Editor vii

Idiano D'Adamo
Adopting a Circular Economy: Current Practices and Future Perspectives
Reprinted from: *Soc. Sci.* **2019**, *8*, 328, doi:10.3390/socsci8120328 1

Idiano D'Adamo
The Profitability of Residential Photovoltaic Systems. A New Scheme of Subsidies Based on the Price of CO_2 in a Developed PV Market
Reprinted from: *Soc. Sci.* **2018**, *7*, 148, doi:10.3390/socsci7090148 7

Mihail Busu
Adopting Circular Economy at the European Union Level and Its Impact on Economic Growth
Reprinted from: *Soc. Sci.* **2019**, *8*, 159, doi:10.3390/socsci8050159 29

Erika Urbánková
The Development of the Health and Social Care Sector in the Regions of the Czech Republic in Comparison with other EU Countries
Reprinted from: *Soc. Sci.* **2019**, *8*, 170, doi:10.3390/socsci8060170 41

Giulia Caruso and Stefano Antonio Gattone
Waste Management Analysis in Developing Countries through Unsupervised Classification of Mixed Data
Reprinted from: *Soc. Sci.* **2019**, *8*, 186, doi:10.3390/socsci8060186 59

Pasquale Marcello Falcone
Tourism-Based Circular Economy in Salento (South Italy): A SWOT-ANP Analysis
Reprinted from: *Soc. Sci.* **2019**, *8*, 216, doi:10.3390/socsci8070216 75

Fernando E. Garcia-Muiña, Rocío González-Sánchez, Anna Maria Ferrari, Lucrezia Volpi, Martina Pini, Cristina Siligardi and Davide Settembre-Blundo
Identifying the Equilibrium Point between Sustainability Goals and Circular Economy Practices in an Industry 4.0 Manufacturing Context Using Eco-Design
Reprinted from: *Soc. Sci.* **2019**, *8*, 241, doi:10.3390/socsci8080241 91

Serdar Türkeli and Martine Schophuizen
Decomposing the Complexity of Value: Integration of Digital Transformation of Education with Circular Economy Transition
Reprinted from: *Soc. Sci.* **2019**, *8*, 243, doi:10.3390/socsci8080243 113

María D. De-Juan-Vigaray and Ana I. Espinosa Seguí
Retailing, Consumers, and Territory: Trends of an Incipient Circular Model
Reprinted from: *Soc. Sci.* **2019**, *8*, 300, doi:10.3390/socsci8110300 135

About the Special Issue Editor

Idiano D'Adamo is a Post-Doctoral Research Fellow at Unitelma Sapienza—University of Rome. He has worked at the University of Sheffield, the National Research Council of Italy, Politecnico di Milano, and University of L'Aquila. In August 2015, he obtained the Elsevier Atlas Prize with a work published in Renewable and Sustainable Energy Reviews. He received two Excellence Review Awards: Waste Management in 2017 and Resources Conservation and Recycling in 2018. During his academic career, Idiano D'Adamo published 57 papers in the Scopus database, receiving 1386 citations and reaching an h-index of 22. His current research interests are bioeconomy, circular economy, renewable energy, sustainability, and waste management.

Editorial

Adopting a Circular Economy: Current Practices and Future Perspectives

Idiano D'Adamo

Department of Law and Economics, Unitelma Sapienza—University of Rome, Viale Regina Elena 295, 00161 Roma, Italy; idiano.dadamo@unitelmasapienza.it

Received: 4 December 2019; Accepted: 6 December 2019; Published: 9 December 2019

Abstract: All scientists, researchers, and citizens are involved in achieving sustainable goals. Their current actions contribute to writing a story for future generations, and interesting perspectives can be narrated based only on a great sense of social responsibility. The literature gives a great deal of attention to the models of a Circular Economy (CE). This topic is multidisciplinary and different sectors are involved in its development. This Special Issue aims to underline the relevance of the CE models in the scientific field and its applications in real contexts in order to achieve sustainability goals.

Keywords: circular economy; social sciences; sustainability

1. Introduction

The Circular Economy (CE) model is able to support sustainable development and has gained attention among policy makers, scholars, and practitioners (Ghisellini et al. 2016). The European Environment Agency has interpreted the CE as the core of a Green Economy perspective that extends the focus from waste and material use to human well-being and ecosystem resilience (see Figure 1) (European Environment Agency 2015). Transitioning to a CE cannot only be used to overlap the linear 'take, make, and dispose' economic model and the Ellen MacArthur Foundation identifies three principles in this regard (The Ellen MacArthur Foundation 2013):

i. "design out waste and pollution".
ii. "keep products and materials in use".
iii. "regenerate natural systems".

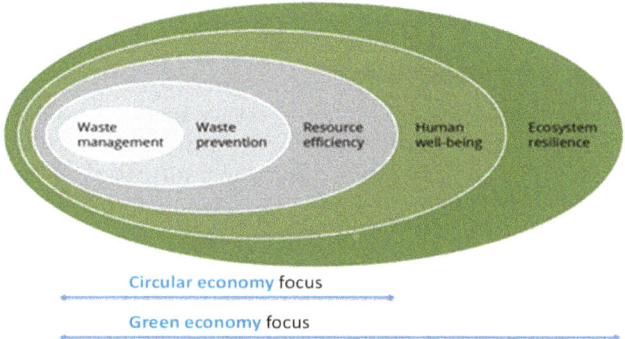

Figure 1. A circular economy and green economy (European Environment Agency 2015).

The World Economic Forum in collaboration with several organizations, such as the Ellen MacArthur Foundation and World Resources Institute, has published a document called Platform for

Accelerating the Circular Economy (PACE) in 2017 with the aim to adopt the principles of CE on the global scale. The waste created by a linear economy damages human health and the environment. Instead, waste that comes from several processes and is inserted into a circular economy provides "beneficial artifacts" for human use (Sikdar 2019). The importance of engineering considerations and design is considered to be vital in order to develop processes in which all parts of a material can be reused/recycled/recovered, thereby minimizing the amount of waste and its dangerousness if it would otherwise end up in landfill (Varbanov and Walmsley 2019).

The conceptualizing of the CE is provided by several works. Some authors provided the following observations. The first that CE is not always associated with the 3R framework (reduce, reuse, recycling) and generally with the waste hierarchy, but instead is often referred to only in the context of recycling. The second that CE is not only a change of the status quo, but also requires a change to a system perspective. The third suggests that the link between CE and sustainable development is weak, as economic prosperity followed by environmental quality are considered, while the impact on social equity is not analyzed. Finally, business models and consumers are evaluated as enablers of CE (Kirchherr et al. 2017).

CE aims to reduce both virgin material inputs and waste outputs by closing resource flow loops. Instead, the goals of sustainability are open-ended. However, the CE is defined as a condition for sustainability, and there are a wide range of complementary strategies that managers and policymakers can adopt (Geissdoerfer et al. 2017). In this way, the value of products and materials is maintained for as long as possible. Consequently, this approach assigns a relevance to material cycles characterized by high value and high quality (Amato et al. 2019). This provides an opportunity to develop an economic model in which both the production phase and the consumption phase are directed towards the protection of the environment (Korhonen et al. 2018). The direct relationship between the CE model and sustainable development is verified by quantitative analysis in which the profitability of the investment project and the reduction of the greenhouse gas emissions are both defined (D'Adamo et al. 2019).

This transition towards circularity requires us to measure its effects. It is possible to estimate a consistent number of indicators and they are specified in function of several criteria, as the levels of CE implementation, the CE loops, the performance, the perspective of circularity, and the degree of transversality (Saidani et al. 2019). The most commonly used methodologies to assess CE are Life Cycle Assessment, Life Cycle Inventory, and Life Cycle Impact Assessment followed by a Multi-Criteria Decision Making approach/fuzzy methods and Design for X (Sassanelli et al. 2019). The final aim is to demonstrate whether restoration and closed-loop product lifecycles are able to reduce waste, minimize the impact of toxic and harmful substances, keep the added value embedded in products and materials, and encourage the use of renewables. In addition, CE plans and targets must be characterized by a "human component", through training and building capacities: an integrative green human resource management framework is provided to support organizations (Jabbour et al. 2019).

2. Form and Contents of the Thematic Issue

Based on these concepts, this Special Issue tries to add new knowledge to the existing literature on the CE. The final aim is to support the adoption of the CE paradigm in companies and organizations around the world. Due to the different sectors and perspectives related to the application of this model, the following papers propose several approaches.

The first work focuses its attention on the photovoltaic source, which represents a vital actor in a transition towards a low-carbon society. The study investigated both environmental and economic performances of photovoltaic systems in a market evaluating different policy scenarios. Subsidies provide a significant increase to profitability, and the reduction of greenhouse gas emissions is calculated as the difference between ones created by an energy mix based on fossil fuels and ones created by photovoltaic plants. Findings show the positive role of a renewable resource towards the link between sustainable development and CE application (D'Adamo 2018).

In a subsequent paper, economic growth was used as the dependent variable, in a function of independent variables, such as the productivity of resources, employment in the production of environmental goods, the recycling rate of municipal waste, market shares of innovative enterprises, and renewable energy uses. The analysis was conducted at the European Union level and the econometric model used demonstrates that the CE factors are relevant indicators of economic growth. Findings of this model permit us to quantify the contribution of each independent variable to the circularity with a primary role played by the productivity of resources (Busu 2019).

The third paper considers the quantitative status of employees in the Health and Social Care sector. Initially, it mentions that this sector fits the theory of unbalanced growth, in which there is has been decreasing productivity calculated based on the gross value added per employee. In the second part of the work, several European countries were examined in terms of their shares of employed persons in professions of this sector, providing specific country clusters. Also the Health and Social Care sector supports the development of circularity, thanks to the availability of technology and cooperation possibilities among all interested parties (Urbánková 2019).

Another work is based on a mixed data, involving continuous and categorical variables and it is aimed to evaluate the performance of solid waste management. This typology of waste represents an opportunity for municipal authorities to maximize the value of materials embedded in these waste and minimize the landfill use. A cluster analysis was implemented on these data. Findings of this work show the relationship of both waste generation and levels of CO_2 emissions with recycling activities and awareness campaigns. Their role has been demonstrated be relevant towards the implementation of good practices of circularity (Caruso and Gattone 2019).

The fifth paper explores the potential development of a second-generation biorefinery in a touristic area trying to integrate waste management, renewable energy, and bio-product production. A Strengths, Weaknesses, Opportunities and Threats Analytical Network Process was used as a model. The results calculated the global priority of each factor and found that social acceptability occupies the first position, followed by excessive bureaucracy, green jobs, lack of long-term planning by governments, and lack of infrastructure technology. Circularity can be supported by some policy strategies, as e-government services, information campaigns, and public infrastructural investments (Falcone 2019).

A subsequent paper evaluates the principles of the CE applied to the manufacturing context. Specifically, a new business model for a ceramic tile manufacturer has been investigated evaluating the impact of an eco-design with a supply system of raw materials. The performance of the company was evaluated while considering the distance of the sources of supply from the factory and relative transport systems. In addition, a recycling process was proposed for the fired waste generated during the production phase. Findings show that eco-design associated with Industry 4.0 Internet of Things technologies can reach the equilibrium point between sustainability and CE (García-Muiña et al. 2019).

Another work investigated the opportunities of the digital transformation of education on CE development to reach the goals of sustainability outcomes. Content analysis and the qualitative meta-synthesis of scientific works referred to digital education for sustainability were used as the methodology. Integrated findings were proposed for capital- and neo-capital-based multiple value formations, for emerging tools and technologies, for micro level interactions among actors and structures, and for macro level interactions among actors, structures, and technologies. As such, the learning of the value embedded in the CE transition requires customizable niches of learning preferences (Türkeli and Schophuizen 2019).

Finally, the last paper concerns the retail distribution and the transition from a linear to an incipient circular retail model. The framework used was based on the Retail Wheel Spins Theory and the Retail Life Cycle. This new circular model can be an opportunity for small entrepreneurs if they are able to intercept the production/selling of products based on a green-image and at the same time, consumers indicate growing interest towards this brand. The theoretical approach suggests that the CE transition can represent a solution to the crisis of the local market favoring the local image of the specific municipalities and provides a contrast to the power of the digital market (De-Juan-Vigaray and Seguí 2019).

3. Concluding Remarks and Further Issues on the Research Agenda

This Special Issue has demonstrated that the field of social sciences is interested in the development of CE models. Its development concerns several sectors and different authors have underlined this as circular economy concept moves towards the final aim of sustainability. The development of a closed loop cycle is a necessary condition to develop a CE model as an alternative to the linear model in order to maintain the value of products and materials for as long as possible. For this motive, the definition of the value must be demonstrated for both the environment and the economy. The presence of these analyses should be associated with the social dimension and the human component.

This editorial suggests some areas of research to investigate in the future:

- the current state of CE. Programs and initiatives in both developing and developed countries.
- future trends of CE. A program of change involving managers, consumers, and politicians.
- the analysis of CE policies. The role of subsidies, penalties, and taxes.
- an assessment of CE. The analysis of potential social opportunities, environmental improvements, and economic advantages.
- measurement of CE. The definition of indicators and the quantification of the circularity of a product.
- rethinking the concept of waste. Needs and opportunities.
- the transition towards a low carbon society. The relationship between the CE models and the use of renewable energies.
- the change on the production side (innovation, efficiency, and efficacy of firms) and on the demand side (the attitudes, behaviors, and practices of consumers).

A strong cooperation between social and technical profiles is a new challenge for all researchers. The End of Life of products attracts a lot of attention and the final output could be the production of technologies suitable for managing this waste and in doing so quantifying both economic and environmental benefits according to the principle of CE.

Funding: This research received no external funding.

Conflicts of Interest: The author declares no conflict of interest.

References

Amato, Alessia, Alessandro Becci, Ionela Birloaga, Ida De Michelis, Francesco Ferella, Valentina Innocenzi, Nicolo Maria Ippolito, Pillar C. Gomez, Francesco Vegliò, and Francesca Beolchini. 2019. Sustainability analysis of innovative technologies for the rare earth elements recovery. *Renewable and Sustainable Energy Reviews* 106: 41–53. [CrossRef]

Busu, Mihail. 2019. Adopting Circular Economy at the European Union Level and Its Impact on Economic Growth. *Social Sciences* 8: 159. [CrossRef]

Caruso, Giulia, and Stefano Antonio Gattone. 2019. Waste Management Analysis in Developing Countries through Unsupervised Classification of Mixed Data. *Social Sciences* 8: 186. [CrossRef]

D'Adamo, Idiano. 2018. The Profitability of Residential Photovoltaic Systems. A New Scheme of Subsidies Based on the Price of CO_2 in a Developed PV Market. *Social Sciences* 7: 148.

D'Adamo, Idiano, Pasquale Marcello Falcone, and Francesco Ferella. 2019. A socio-economic analysis of biomethane in the transport sector: The case of Italy. *Waste Management* 95: 102–15. [CrossRef] [PubMed]

De-Juan-Vigaray, Maria D., and Ana I. Espinosa Seguí. 2019. Retailing, Consumers, and Territory: Trends of an Incipient Circular Model. *Social Sciences* 8: 300. [CrossRef]

European Environment Agency. 2015. *Circular Economy in Europe—Developing the Knowledge Base*. Luxembourg: Publications Office of the European Union.

Falcone, Pasquale Marcello. 2019. Tourism-Based Circular Economy in Salento (South Italy): A SWOT-ANP Analysis. *Social Sciences* 8: 216. [CrossRef]

Garcia-Muiña, Fernando E., Rocio González-Sánchez, Anna Maria Ferrari, Lucrezia Volpi, Martina Pini, Cristina Siligardi, and Davide Settembre-Blundo. 2019. Identifying the Equilibrium Point between Sustainability Goals and Circular Economy Practices in an Industry 4.0 Manufacturing Context Using Eco-Design. *Social Sciences* 8: 241.

Geissdoerfer, Martin, Paulo Savaget, Nancy Bocken, and Erik Jan Hultink. 2017. The Circular Economy—A new sustainability paradigm? *Journal of Cleaner Production* 143: 757–68. [CrossRef]

Ghisellini, Patrizia, Catia Cialani, and Sergio Ulgiati. 2016. A review on circular economy: The expected transition to a balanced interplay of environmental and economic systems. *Journal of Cleaner Production* 114: 11–32. [CrossRef]

Jabbour, Chiappetta, Charbel Jose, Josiph Sarkis, Ana Betriz Lopes de Sousa Jabbour, Douglas William Scott Renwick, Sanjay Kumar Singh, Oksana Grebinevych, Isak Kruglianskas, and Moacir Godinho Filho. 2019. Who is in charge? A review and a research agenda on the 'human side' of the circular economy. *Journal of Cleaner Production* 222: 793–801. [CrossRef]

Kirchherr, Julian, Denise Reike, and Marko Hekkert. 2017. Conceptualizing the circular economy: An analysis of 114 definitions. *Resources Conservation and Recycling* 127: 221–32. [CrossRef]

Korhonen, Jouni, Antero Honkasalo, and Jyri Seppälä. 2018. Circular Economy: The Concept and its Limitations. *Ecological Economics* 143: 37–46. [CrossRef]

Saidani, M., B. Yannou, Y. Leroy, F. Cluzel, and A. Kendall. 2019. A taxonomy of circular economy indicators. *Journal of Cleaner Production* 207: 542–59. [CrossRef]

Sassanelli, Claudio, Paolo Rosa, Roberto Rocca, and Sergio Terzi. 2019. Circular economy performance assessment methods: A systematic literature review. *Journal of Cleaner Production* 229: 440–53. [CrossRef]

Sikdar, Subhas. 2019. Circular economy: Is there anything new in this concept? *Clean Technologies and Environmental Policy* 21: 1173–75. [CrossRef]

The Ellen MacArthur Foundation. 2013. Towards the Circular Economy: Economic and Business Rationale for an Accelerated Transition. Available online: https://www.ellenmacarthurfoundation.org/assets/downloads/publications/Ellen-MacArthur-Foundation-Towards-the-Circular-Economy-vol.1.pdf (accessed on 9 December 2019).

Türkeli, Serdar, and Martine Schophuizen. 2019. Decomposing the Complexity of Value: Integration of Digital Transformation of Education with Circular Economy Transition. *Social Sciences* 8: 243. [CrossRef]

Urbánková, Erika. 2019. The Development of the Health and Social Care Sector in the Regions of the Czech Republic in Comparison with other EU Countries. *Social Sciences* 8: 170. [CrossRef]

Varbanov, Petar Sabev, and Timothy Gordon Walmsley. 2019. Circular economy and engineering concepts for technology and policy development. *Clean Technologies and Environmental Policy* 21: 479–80. [CrossRef]

© 2019 by the author. Licensee MDPI, Basel, Switzerland. This article is an open access article distributed under the terms and conditions of the Creative Commons Attribution (CC BY) license (http://creativecommons.org/licenses/by/4.0/).

Article

The Profitability of Residential Photovoltaic Systems. A New Scheme of Subsidies Based on the Price of CO_2 in a Developed PV Market

Idiano D'Adamo

Department of Industrial and Information Engineering and Economics, University of L'Aquila, 67100 L'Aquila, Italy; idiano.dadamo@univaq.it

Received: 4 August 2018; Accepted: 27 August 2018; Published: 31 August 2018

Abstract: Photovoltaic (PV) resource drives the clean global economy of the future. Its sustainability is widely confirmed in literature, however some countries present a growth very low in the last years. A new policy proposal is examined in this work. It aims to stimulate a new diffusion of PV plants in mature markets (e.g., Italy) regarding residential consumers. The subsidy is given to the amount of energy produced by PV plant for a period of 20 years (equal to its lifetime) and its value is calculated according to the scheme of European Emissions Trading System (EU ETS). Discounted Cash Flow (DCF) is used as economic method and two indexes are proposed: Net Present Value (NPV) and Discounted Payback Time (DPBT). The baseline case studies vary in function of two variables; (i) the share of self-consumption (30%, 40% and 50%) and (ii) the price of emissions avoided (10, 35 and 70 € per ton of CO_2eq). Results confirms the environmental advantages of PV sources as alternative to the use of fossil fuels (685 gCO_2eq/kWh) and economic opportunities are verified in several scenarios (from 48 €/kW to 1357 €/kW). In particular, the profitability of PV systems is greater with a subsidized rate of fiscal deduction of 50% in comparison to subsidies with a value of carbon dioxide lower than 18.50 €/tCO_2eq.

Keywords: CO_2 emissions; economic analysis; photovoltaic; subsidies

1. Introduction

Social Sciences aims to integrate considerations regarding the sustainability of humanity (Lin 2012). The global warming is one the most important hazards for the Earth's future and the use of renewable energy sources (RES) is a valid solution to stop their adverse influences on human life (Saavedra et al. 2018).

Global energy demand increased by 2.1% in 2017 and also, global energy-related CO_2 emissions grew by 1.4% in 2017 (IEA 2015). Recently, the whole energy sector changes towards the use of low-carbon applications. Renewable energy (RE) power generating capacity is equal to 2195 GW in 2017 (+8.8% than previous year). This electricity transition is driven by increases in installed capacity of solar PV (+99 GW with an increase of 32.7% than 2016) and wind power (+52 GW with an increase of 10.7% than 2016)—Figure 1 (REN21 2018).

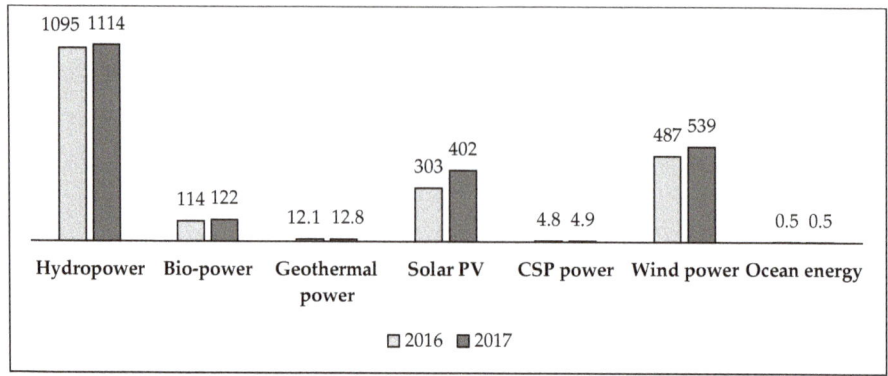

Figure 1. Cumulative global renewable power capacity. Data expressed in GW (REN21 2018).

Economic growth is typically coupled with the use of energy consumption (Sun et al. 2018). However, the energy consumption is usually linked to a great level of emissions and pollutions. This effect is significantly reduced when the green electricity is used (Sampaio and González 2017). In addition, two actions push towards more effective future global initiatives. The first regards strategies that engage all political parties, the second aims to educate individuals on climate change (Dadural and Reznikov 2018). At the same time, residential energy consumption can be improved not only through adequate technological solutions but also with a behavior more eco-friendly to citizens (Escoto Castillo and Peña 2017).

PV sources can play a key role in this energy transition for the global energy supply (Breyer et al. 2017). Solar PV is a mature technology suitable for both small and large scale applications. It is a clean energy according to the principle of sustainability (Hosenuzzaman et al. 2015; Khan and Arsalan 2016). Solar PV power capacity is equal to 402 GW in 2017 and it is concentrated in a short list of countries. In fact, about 86% of this power is installed in 10 countries with a role predominant of China (Figure 2). China (53.1 GW), United States (10.6 GW) and India (9.1 GW) represent the first three countries of solar PV power installed in 2017 (REN21 2018).

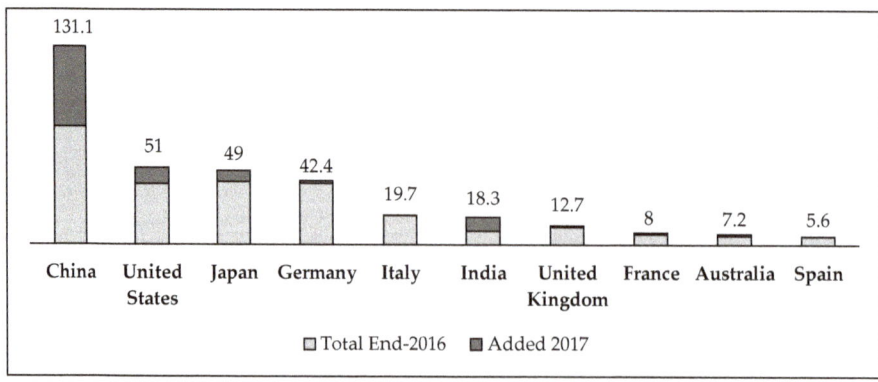

Figure 2. Cumulative solar PV power capacity in 2017. Data expressed in GW. Top 10 countries (REN21 2018).

The Feed-in-Tariff (FIT) scheme has encouraged investors to be involved in RE production worldwide. Large energy providers offer long-term contracts to smaller-scale RE producers to sell their green energy to the market under a fixed tariff above the market rate (Pyrgou et al. 2016; Tanaka et al. 2017). The policy subsidy has determined the development of PV source with the aim to tackle the climate change. At the same time, the guaranteed security of tariffs, defined in a FIT scheme, has driven several investors to choose this resource (Avril et al. 2012; Strupeit and Palm 2016). In addition, it has determined an improvement of the technology, a reduction of costs and an increase of know-how of firms (Baur and Uriona 2018).

The economic feasibility of PV plants is well analysed in literature. Residential applications represent a typical case-study (Lee et al. 2017; Comello and Reichelstein 2017). The key-parameter of profitability depends by the typology of the market in residential PV systems: subsidies and the share of self-consumption are the main variable in developing and developed markets, respectively (Cucchiella et al. 2017a).

From environmental side, the greenhouse gas (GHG) emissions produced by PV systems are estimated equal to 29–35 gCO_2eq/kWh (Fthenakis et al. 2008). Literature analysis presents a variety of approaches to calculate GHG emissions. Consequently, there is a wide variety in the evaluation of this value: for example some authors propose 20–25 gCO_2eq/kWh (Louwen et al. 2016), other 60.1–87.3 gCO_2eq/kWh (Hou et al. 2016). However, all studies converge to define that this environmental effect is widely balanced by the reduction of GHG emissions determined by the use of PV resource as alternative to fossil fuels. Assuming a lifetime of PV plant equal to 20 year, the environmental advantage is quantified equal to 21 tCO_2eq per kW installed (Cucchiella et al. 2016). Another work has calculated a reduction of about 742.7 gCO_2eq/kWh. It considers 37.3 gCO_2eq/kWh and 780 gCO_2eq/kWh for PV and coal resources, respectively (Mauleón 2017).

A review of CO_2 price with government subsidy through FIT scheme is analysed for European countries (Bakhtyar et al. 2017). The evaluation of PV systems under carbon market is proposed also in Chinese context (Tian et al. 2017). A low carbon tax is able to finance the investment in PV plants (Mauleón 2017). The economic evaluation of PV systems is required for the development of the sector also in a market developed (Cucchiella et al. 2017b). A new research can try to consider policy, environmental and economic aspects. This work proposes the economic impact of a residential PV plant and a small size equal to 3 kW located in Italy is considered. The idea is to implement a new policy of subsidies for residential consumers that implemented PV systems. The subsidy is given to the amount of energy produced and its value is calculated according to the reduction of CO_2 emissions.

The paper is organised as follows. Section 2 presents a literature review concerning the mechanisms of market of CO_2. An economic model based on DCF is proposed in Section 3. Starting by input data, NPV and DPBT are used to evaluate the economic performance of PV systems considering several scenarios (Section 4). Section 5 presents some concluding remarks.

2. Literature Review

The European Union (EU) launched the EU ETS to fight global warming in 2005. EU ETS covers around 11,000 power stations and industrial plants. The inspiring principle of EU ETS is to give firms an incentive to move towards less fossil-fuel intensive production. It works on the 'cap and trade' principle. The emission allowance (EUA) allows the firms to emit one tonne of CO_2 and each of them has assigned a limit of CO_2 emissions (cap). The following year, a defined number of EUAs must be returned. If this number is lower than the assigned cap, the firm has the opportunity to sell EUAs (trade). When, instead, it is greater the firm must buy the missing shares. Alternately, heavy fines are provided. The limit is reduced over time so that total emissions decrease (European Commision 2016).

Several works have considered the European context. Energy prices are considered by some authors as the main driver of carbon price because power generators can use several fuel inputs (Christiansen et al. 2005; Convery and Redmond 2007). Other works have underlined the relevance of other critical variables as weather conditions, policy and regulatory issues and economy activities. Prices vary to uncontrollable temperatures changes during colder events (Alberola et al. 2008). At the same time, institutional strategies have a direct impact (Aatola et al. 2013). In fact, during the First Phase of EU ETS coal and gas prices have influenced CO_2 prices, while electricity price has played a role more during the Second Phase (Keppler and Mansanet-Bataller 2010). Foreign direct investment (FDI) increase carbon emissions in the host country influencing the carbon price (Doytch and Uctum 2016).

The market instrument of CO_2 ETS is been implemented also in several Chinese regions and it is regulated by the government (Yang et al. 2017). The analysis of market highlights that the carbon price is closely linked to the supply and demand of carbon allowance. The supply is determined by Government policies, while the demand is determined by the regional economic pattern and energy structure (Yang et al. 2018). The development of an ETS is more complex in a vast country with regional differences (Böhringer et al. 2014). Other international initiatives to tackle the increase of CO_2 emissions are California cap-and-trade program (Olson et al. 2016), cap-and-trade programs of the Republic of Korea (Park and Hong 2014). A comparative among several programs is investigated and EU ETS is the main cornerstone to combat climate change (Xiong et al. 2017).

However, several works have identified the criticism of EU ETS. Three limits are identified: (i) it is not an attractive market for its economic added value, (ii) it is not able to maintain the carbon price sufficiently high and (iii) it has no reduced significantly the overall emissions (Gerbeti 2017). In particular, EU ETS had not encouraged green investments (Segura et al. 2018) and its ineffectiveness is substantiated in times of economic crisis (Vlachou and Pantelias 2017). Another work defines that EU ETS lacks fairness on both effectiveness and the distribution of the duties involved in climate change (Dirix et al. 2015). The risk of carbon leakage is extremely high for energy-intensive industries. Some firms can transfer their production in countries with lower emission constraints (Gerbeti 2018). This work does not aim to define a judgement on EU ETS. It is based on the approach that the emissions must be quantified in economic terms and considering the European context, in this moment EU ETS represents the main reference.

Literature review has covered mainly the first two phases of EU ETS. The main mechanism was free allocation based on past emissions. Since 2013, auctioning is the default method of allocating emission allowances (Cai and Pan 2017). The accurate prediction of carbon prices is an information useful for carbon traders, brokers and firms, who can use this information to manage their portfolios. This data is necessary also for policy makers, who have inputs on marginal abatement costs adjusting the emission cap (Zhao et al. 2018).

The development of carbon trading aims to tackle the climate change, to improve the energy system, to promote energy-saving and emission-reduction (ESER) system and to accelerate the transformation of economic growth (Fang et al. 2018b). The government control is a sensitive parameter in carbon trading system. In fact, policy measures can accelerate its development reaching the peak value of carbon emissions in short terms, but the effect can be also negative in specific economic periods. The equilibrium between demand and supply requires generally a run-in period to achieve balance (Fang et al. 2018a).

Carbon price is a tool for scientists to reduce global warming. The value indicated by several authors varies in a significant way. Nationally efficient CO_2 prices are referred to domestic environmental benefits per ton of CO_2 reduction. For example, it is equal to 63 \$/tCO_2 and 57.5 \$/tCO_2 in USA and China in 2010, respectively. A greater difference is instead found for 2013 between Europe (below 10 \$/tCO_2) and USA (35 \$/tCO_2) (Parry et al. 2015). Another work has calculated a global carbon price in order to estimate the annual transfer payments that would be required to compensate the damages linked to the emissions. It is equal to 35 \$/tCO_2 (Landis and Bernauer 2012). Other authors quantified the economic advantages linked to the technological solutions able to capture CO_2 emissions. Benefits are evaluated considering a price of 13 \$/tCO_2 (Ogland-Hand et al. 2017). The substitution of fossil fuels with a renewable resource (wind) is evaluated in Chinese context. Carbon price varies from 233 CNY/tCO_2 to 251 CNY/tCO_2 and it is higher than real markets because a high proportion of free allowances is used (Lin and Chen 2018).

A group of economists has defined that about 75% of emissions regulated by carbon pricing are covered by a price below 10 €/tCO$_2$ in 2017. This price is considered too low in order to support the low carbon transition (Metivier et al. 2017). There are other studies (Gerbeti 2016) that claim to economically enhance the CO$_2$ contained in the goods, representing it as a raw material of industrial production processes.

The effective carbon rate (ECR) is the sum of carbon taxes, specific taxes on energy use and tradable emission permit prices. The OECD has estimated the ECR for 41 countries. ECR is assumed equal to 30 €/tCO$_2$ (OECD 2016). This value is lower than other studies: 50 €/tCO$_2$ (Alberici et al. 2014) and 50 $/tCO$_2$ (Smith and Braathen 2015).

A recent report of the High-Level Commission on Carbon Prices guided by Stiglitz and Stern has defined relevant several indications for the future. From one side, a consistent quantity of emissions are not covered by a carbon price and from the other side, about three quarters of the emissions have a price lower than 10 $/tCO$_2$. The Nationally Determined Contributions (NDCs) for 2030 associated with the Paris Agreement are not suitable to achieve the Paris target of "well below 2 °C." This target could be reach using a price from 40 $/tCO$_2$ to 80 $/tCO$_2$ by 2020 and from 50 $/tCO$_2$ to 100 $/tCO$_2$ by 2030. In fact, the use of carbon pricing must be considered also non-climate benefits, for example access to modern energy, the health of ecosystems and improvements in air pollution and congestion (Stiglitz et al. 2017).

Some authors have identified the value of certified emission reduction equal to 20 CNY/ tCO$_2$ and it is applied a case study of PV systems. Their results define that firms have not benefits until carbon price does not exceed 38 CNY/tCO$_2$ (Tian et al. 2017). A comprehensive review has identified the social cost of carbon. Its minimum value is equal to 6.1 €/tCO$_2$ (Isacs et al. 2016). The value of CO$_2$ emissions is strictly linked to possible economic downturns and also to the volatility of energy prices in an organized market, as EU ETS (Mauleón 2017). The substitute price of avoiding CO$_2$ emission (SPAC) is calculated for each technology and country in Europe. Values obtained are extremely far from market prices (Bakhtyar et al. 2017).

3. Materials and Methods

The methodology used in this paper is based on several steps:

1. The definition of emissions avoided using PV resource as alternative to the fossil fuels.
2. The evaluation of CO$_2$eq emissions price.
3. The policy proposal.
4. The economic model.
5. The presentation of case studies.
6. Input data.

3.1. The Reduction in the Emissions of Carbon Dioxide

From environmental side, there is a reduction in the Emissions of Carbon Dioxide (RECD) when the energy is produced using a PV system compared to the use of fossil fuels. Starting by a hypothetical energy mix composed only by fossil fuels and considering results of literature review regarding GHG emissions from fossil fuels, the value of emissions released by a mix of fossil fuels (ECD$_{FF}$) is calculated—Equation (1). The definition of emissions released by PV source (ECD$_{PV}$) is defined considering also in this case the results of literature review regarding GHG emissions from this resource. In this way, it is possible to calculate RECD as difference between ECD$_{FF}$ and ECD$_{PV}$—Equation (2).

$$ECD_{FF} = ECD_{OIL} \times PEM_{OIL} + ECD_{COAL} \times PEM_{COAL} + ECD_{GAS} \times PEM_{GAS} \quad (1)$$

$$RECD = ECD_{FF} - ECD_{PV} \quad (2)$$

in which ECD$_{OIL}$ = emissions of carbon dioxide released by oil, ECD$_{COAL}$ = emissions of carbon dioxide released by coal, ECD$_{GAS}$ = emissions of carbon dioxide released by natural gas, PEM$_{OIL}$ = percentage in energy mix of oil, PEM$_{COAL}$ = percentage in energy mix of coal and PEM$_{GAS}$ = percentage in energy mix of natural gas.

Figure 3 reports several values concerning the Life Cycle Analysis of GHG emissions from electricity generation technologies. The difference between fossil fuels and RES is extremely significant. In this study an average value obtained by values reported in Figure 3 is chosen for the fossil fuels: ECD$_{OIL}$ = 824 gCO$_2$eq/kWh, ECD$_{COAL}$ = 1149 gCO$_2$eq/kWh and ECD$_{GAS}$ = 568 gCO$_2$eq/kWh.

Regarding the emissions of PV systems, values reported in Figure 3 vary from 5 to 92 gCO$_2$eq/kWh, while ones reported in Section 1 from 20 to 87.3 gCO$_2$eq/kWh. PV source is the core of this work and for this motive, other studies are proposed in order to choose an appropriate value: 15–76 gCO$_2$eq/kWh (Bravi et al. 2011), 10.5–50 gCO$_2$eq/kWh (Peng et al. 2013), 13–39 gCO$_2$eq/kWh (Fthenakis and Kim 2013) and 49 gCO$_2$eq/kWh (Cucchiella et al. 2017a). ECD$_{PV}$ = 42 gCO$_2$eq/kWh is the value hypothesized and it is obtained as average value of all studies examined in this work.

The energy report in Italy underline a growth of natural gas occupying a leadership position with a share of 36.5%. The oil continues to decrease (about 34%) with a reduction of ten points in comparison to ten years ago. RES has a share of 19% with a decrease of hydropower and an increase of solar energy and wind. However, it is far by the maximum value (21%) reached in 2014 (ENEA 2018).

In order to evaluate the mix of fossil fuels is used the approach proposed by (Cucchiella et al. 2017a). Energy portfolio is calculated at net of renewables and imports. The following values are obtained for 2017 year: PEM$_{GAS}$ = 48%, PEM$_{OIL}$ = 44% and PEM$_{COAL}$ = 8%. In comparison to the previous year, there is a difference. In fact, the percentage of both gas and oil is equal to 45.5%, while one of coal is 9% in 2016.

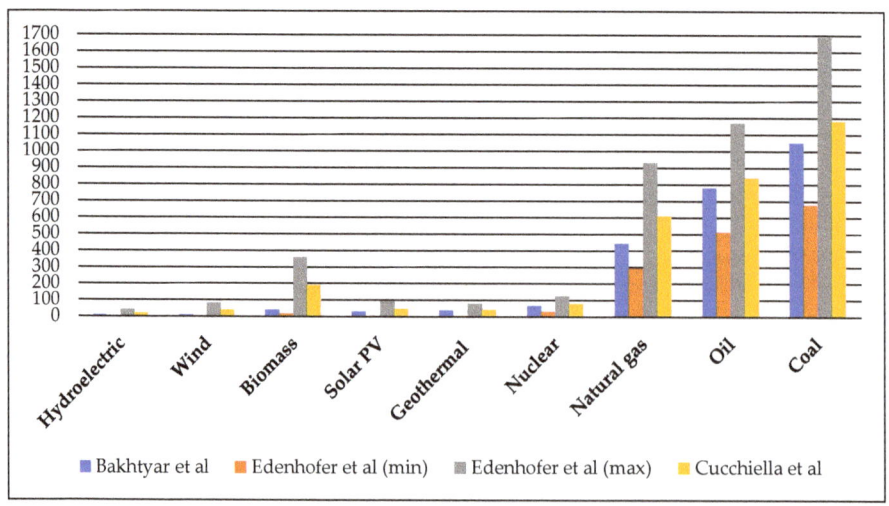

Figure 3. Emissions of carbon dioxide of energy sources. Data expressed in gCO$_2$eq/kWh (Cucchiella et al. 2017a; Bakhtyar et al. 2017; Edenhofer et al. 2012).

3.2. The Price of CO$_2$eq Emissions

Section 2 has underlined that the carbon price is characterized by a great variability. For this motive, the trend of EU ETS is examined during the last year (from 26 July 2017 to 26 July 2018). A value for each month is reported in Figure 4. There is a significant growth in the last year (from 4.84 €/tCO$_2$eq to 16.99 €/tCO$_2$eq). A maximum value equal to 17.45 €/tCO$_2$eq is registered on 24 July 2018.

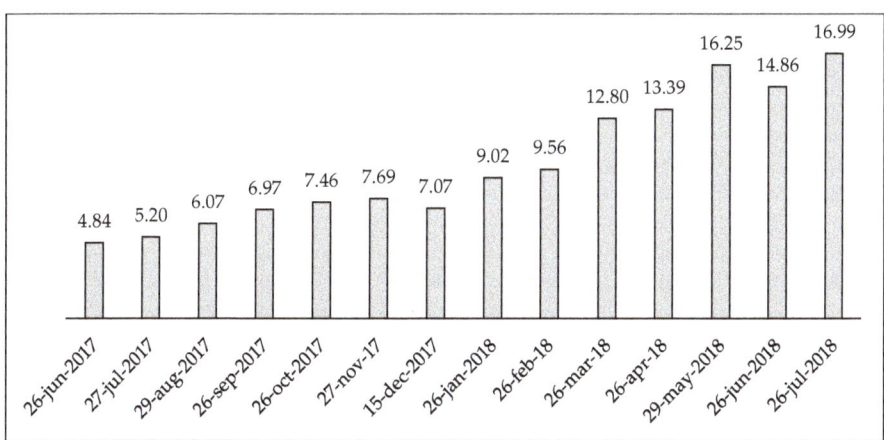

Figure 4. CO$_2$ European emission allowances. Data expressed in €/EUA (Markets Insider 2018).

However, it is opportune to underline as the prices of the last year are following many legislative changes in on-going process of the scheme. Certified emission reductions (CERs) and emission reduction units (ERUs) were effective until 2009. The absence of an agreement post-Kyoto has determined a significant reduction of CO$_2$ price equal to about 14 €/tCO$_2$eq and 17 €/tCO$_2$eq for ERUs and CERs respectively (Gerbeti 2017).

The price of CO$_2$ emissions (P_{CD}) is assumed equal to 10 €/tCO$_2$eq and this choice is assumed according to two motivations. The first regards that this value is the average value reported in Figure 4 and the second concerns the literature review proposed in Section 2, in which this value is often proposed by authors. Literature review and real markets have presented different values in several case studies. According also to the variability of this price, it is opportune to conduct a sensitivity analysis on this variable in order to image future scenarios. Consequently, three scenarios are considered in this work:

1. Low price of CO$_2$ emissions (Low P_{CD}), in which P_{CD} is equal to 10 €/tCO$_2$eq. In fact, it is the baseline value, but at the same time several authors have underlined that this value is not appropriate for a transition towards a society low carbon.
2. Moderate price of CO$_2$ emissions (Moderate P_{CD}), in which P_{CD} is equal to 35 €/tCO$_2$eq. This value represents the minimum value proposed by the report of the High-Level Commission on Carbon Prices.
3. High price of CO$_2$ emissions (High P_{CD}), in which P_{CD} is equal to 70 €/tCO$_2$eq. This value represents the maximum value proposed by the report of the High-Level Commission on Carbon Prices.

3.3. The Policy Proposal

Figure 1 has underlined that Italy occupies the fifth position in the ranking of PV power installed. In the last year, only 0.4 GW are installed. This result is negative, in fact among top ten countries only Spain has registered a lower value (+0.1 GW) (REN21 2018). The definition of negativity is given according to the environmental advantages linked to the use of solar resource.

The tradition tools of subsidies, as Feed-in-Premium and FIT, are no longer provided in this country. The policy choices support the development of PV residential sector through a 50% tax deduction in substitution of the typical value of 36%. The deduction is divided into ten equal yearly amounts.

This work try to propose a new tool of subsidy according to both Paris Agreement and EU ETS. The economic support is given for a period of 20 years, equal to the lifetime of PV plants. The initial assumption defines as all consumers employed to reduce carbon dioxide emission levels can receive an economic contribution. These funds are paid by operators who produce a level of pollutants greater than the value allowed. A comparison with existing literature is not possible, in fact for the first time this idea is applied to PV systems in residential applications.

The unitary value of subsidies (SUB_{PV}) is obtained multiplying three factors: (i) the amount of reduction of CO_2eq emissions substituting the fossil fuels with the production of energy by PV system, (ii) the price of CO_2eq emissions and (iii) the amount of energy produced by PV system during its lifetime—Equation (3). The energy produced is calculated in function of several variables: average annual insolation (t_r), optimum angle of tilt (k_f), module efficiency (η_m), balance of system efficiency (η_{bos}), active surface (A_{cell}), nominal power of a PV module (P_f) and number of PV modules to be installed (η_f)—Equation (4). The value of SUB_{PV} varies during the lifetime in function of reduction of energy produced by PV system because a decrease of efficiency of PV system (dE_f) is considered—Equation (5).

$$SUB_{PV,t} = RECD \times P_{CD} \times E_{out,t} \qquad (3)$$

$$E_{Out,t} = t_r \times K_f \times \eta_m \times \eta_{bos} \times A_{cell} \times P_f \times \eta_f \qquad (4)$$

$$E_{out,t+1} = E_{out,t} \times (1 - dE_f) \qquad (5)$$

In an objective context, the value of SUB_{PV} should vary also in function of RECD. In fact, if ECD_{PV} can be assumed fixed for an operating PV plant, ECD_{FF} varies in function of the energy mix. For example, this value is equal to 727 gCO_2eq/kWh in 2017 while a value greater is obtained in 2016 (737 gCO_2eq/kWh). At the same time, the value of SUB_{PV} should vary also in function of P_{CD}. In fact, this value changes in according to both supply and demand of CO_2. For example, there is difference of about 12 € ton of CO_2eq during the period analysed in Figure 4. This assumption is justified by Section 1, in which the variability of subsidy is perceived as an issue by investors.

The final purpose has a nature not speculative and consequently, it is possible to fix the value of P_{CD} during all lifetime of PV system according to the principle used in FIT scheme.

3.4. The Economic Model

DCF analysis is a method of valuing a project using the concepts of the time value of money. It is based on an incremental approach, in which cash inflows and outflows are considered and a cost opportunity of capital is applied to aggregate several cash flows.

NPV and DPBT are the financial indexes proposed in this work. NPV is the sum of present values of individual cash flows—Equation (6). DPBT is the number of years needed to balance cumulative discounted cash flows and the initial investment—Equation (7) (Cucchiella et al. 2017a).

Four items are hypothesized as revenues: (i) fiscal deduction, (ii) saving energy through internal consumption, (iii) selling energy not used for internal consumption and (iv) subsidies—Equations (8) and (9). Six items are considered as costs: (i) investment, (ii) maintenance, (iii) assurance, (iv) taxes, (v) replacement of inverter and (vi) general—Equations (10) and (11). The mathematical reference model used in a previous research is considered (Cucchiella et al. 2017a) and a new item of revenue (subsidies) is added. The novelty of the work consists also in the calculation of this new value. The model is reported below:

$$NPV = DCI - DCO \qquad (6)$$

$$\sum_{t=0}^{DPBT} (CI_t - CO_t)/(1+r)^t = 0 \qquad (7)$$

$$DCI = \sum_{t=1}^{N}(\omega_{self,c} \times E_{Out,t} \times p_t^c + \omega_{sold} \times E_{Out,t} \times p_t^s)/(1+r)^t + \sum_{t=1}^{N_{TaxD}}((C_{inv}/N_{TaxD}) \times TaxD_u)/(1+r)^t + \sum_{t=1}^{N}(RECD \times P_{CD} \times E_{Out,t})/(1+r)^t \quad (8)$$

$$p_{t+1}^c = p_t^c \times (1+inf_{el}); p_{t+1}^s = p_t^s \times (1+inf_{el}) \quad (9)$$

$$DCO = \sum_{t=0}^{N_{debt}-1}(C_{inv}/N_{debt} + (C_{inv} - C_{lcs,t}) \times r_d)/(1+r)^t + \sum_{t=1}^{N}(P_{Cm} \times C_{inv} \times (1+inf) + P_{Cass} \times C_{inv} \times (1+inf) + SP_{el,t} \times P_{Ctax})/(1+r)^t + (P_{Ci} \times C_{inv})/(1+r)^{10} + C_{ae} \quad (10)$$

$$C_{inv} = C_{inv,unit} \times (1+Vat) \times P_f \times \eta_f \quad (11)$$

in which DCI = discounted cash inflow, DCO = discounted cash outflow, CI = cash inflow, CO = cash outflow, r = cost opportunity of capital, t = time period, N = lifetime of a PV system, $\omega_{self,c}$ = percentage of energy self-consumption, ω_{sold} = percentage of the produced energy sold to the grid, p^c = electricity purchase price, p^s = electricity sales price, C_{inv} = total investment cost, N_{TaxD} = period of tax deduction, $TaxD_u$ = unitary tax deduction, inf_{el} = rate of energy inflation, N_{debt} = period of loan, C_{lcs} = loan capital share cost, r_d = interest rate on a loan, P_{Cm} = percentage of maintenance cost, inf = rate of inflation, P_{Cass} = percentage of assurance cost, SP_{el} = sale of energy, P_{Ctax} = percentage of taxes cost, P_{Ci} = percentage of inverter cost, C_{ae} = administrative and electrical connection cost, $C_{inv,unit}$ = unitary investment cost and Vat = value added tax.

3.5. The Presentation of Case Studies

Subsidies have played a key-role in the development of PV sector. As defined in Section 1, this work aims to propose an economic analysis of PV systems in residential applications. For this motive, a plant size equal to 3 kW is considered.

This work try to evaluate the impact of subsidies on the profitability of PV systems, consequently several scenarios can be analysed:

1. Scenario "Fiscal Deduction 36%", in which subsidies are not provided and Fiscal Deduction has a standard value ($TaxD_u$ = 36%).
2. Scenario "Fiscal Deduction 50%", in which subsidies are not provided (P_{CD} = 0 €/tCO_2eq) and Fiscal Deduction is subsidized ($TaxD_u$ = 50%).
3. Scenario "Subsidies Low P_{CD}", in which subsidies are provided with a low value of P_{CD} and Fiscal Deduction has a standard value of 36%.
4. Scenario "Subsidies Moderate P_{CD}", in which subsidies are provided with a moderate value of P_{CD} and Fiscal Deduction has a standard value of 36%.
5. Scenario "Subsidies High P_{CD}", in which subsidies are provided with a high value of P_{CD} and Fiscal Deduction has a standard value of 36%.

3.6. Input Data

The transformation of both cash inflows and outflows in discounted values requires the use of a cost opportunity of capital. This variable measures the return coming from an alternative project, which has the same risk level. It is hypothesized equal to 5%. The time period of cash flows is defined by the lifetime of PV plant, which is assumed equal to 20 years. PV plant is located in a central region (1450 kWh/m^2 × year) and investment costs are covered by third party funds. The share of self-consumption is the harmonization between demanded and produced energy. This variable assumes a key-role in the economic evaluation and for this motive three scenarios characterized by different values are considered (Cucchiella et al. 2017a):

1. Scenario "Self-consumption 30%", in which the investor uses the 30% of energy produced for internal uses and the remaining share is sold to the market.
2. Scenario "Self-consumption 40%", in which $\omega_{self,c}$ and ω_{sold} are equal to 40% and 60%, respectively.
3. Scenario "Self-consumption 50%", in which the share of self-consumption is equal to 50%.

Other economic inputs useful to develop the economic model presented in the previous sub-section are proposed in Table 1.

Table 1. Economic inputs (Cucchiella et al. 2017b; Orioli et al. 2016).

Variable	Value	Variable	Value
A_{cell}	7 m²/kWp	p^s	5.5 cent€/kWh
C_{ae}	250 €	P_{Cass}	0.4%
$C_{inv,unit}$	1900 €/kW	P_{Ci}	15%
dE_f	0.7%	P_{Cm}	1%
inf	2%	P_{Ctax}	43.5%
$\inf_{el}$	1.5%	P_f	function of S
k_f	1.13	r	5%
N	20 y	r_d	3%
N_{debt}	15 y	S	3 kW
N_{TaxD}	10 y	t_r	1450 kWh/m² × year
η_{bos}	85%	$TaxD_u$	36–50%
η_f	function of S	$w_{self,c}$	30–50%
η_m	16%	w_{sold}	50–70%
p^c	19 cent€/kWh	Vat	10%

4. Results

The first step is represented by the calculation of RECD. This value is reported in Equation (12) and it is applied also for the following years of lifetime of PV systems. Currently, there are no robust estimates on the future energy mix. However, alternative scenarios concerning this variable will be examined in the following section.

$$\text{RECD} = (842 \times 0.44 + 114 \times 0.08 + 568 \times 0.48) - 42 = 685 \text{ gCO}_2\text{eq/kWh} \tag{12}$$

The following step is the economic quantification of reduction of carbon dioxide. According to Equation (4) and input data reported in Table 1, $E_{Out,1}$ is equal to 4680 kWh/year during the first year. Consequently the unitary value of subsidies is reported in Equations (13)–(15) according to the single value of P_{CD}.

$$\text{SUB}_{PV,1} = 685 \times 10 \times 4680 = 32 \text{ €/year} \quad \text{Subsidies Low } P_{CD} \tag{13}$$

$$\text{SUB}_{PV,1} = 685 \times 35 \times 4680 = 112 \text{ €/year} \quad \text{Subsidies Moderate } P_{CD} \tag{14}$$

$$\text{SUB}_{PV,1} = 685 \times 70 \times 4680 = 224 \text{ €/year} \quad \text{Subsidies High } P_{CD} \tag{15}$$

The results of economic feasibility are subdivided as follows:

1. Baseline scenarios.
2. The distribution of revenues.
3. Alternative scenarios.
4. Discussions and policy implications.

4.1. Baseline Scenarios

The profitability of a 3 kW PV plant is evaluated in this work. The baseline scenario is composed by fifteen case studies obtained multiplying three scenarios linked to consumer choices and five scenarios related to political decisions. Two distinct indexes are proposed, because NPV quantifies the amount of money generated by PV investment (Table 2), while DPBT gives an information concerning the number of years in which the investment is recovered (Table 3).

Table 2. NPV in baseline scenario. Data expressed in €.

Scenarios	Self-Consumption 30%	Self-Consumption 40%	Self-Consumption 50%
Fiscal Deduction 36%	−533	455	1443
Fiscal Deduction 50%	145	1133	2121
Subsidies Low P_{CD}	−158	830	1819
Subsidies Moderate P_{CD}	780	1769	2757
Subsidies High P_{CD}	2094	3082	4070

Table 3. DPBT in baseline scenario. Data expressed in years.

Scenarios	Self-Consumption 30%	Self-Consumption 40%	Self-Consumption 50%
Fiscal Deduction 36%	>20	18	13
Fiscal Deduction 50%	19	15	5
Subsidies Low P_{CD}	>20	16	6
Subsidies Moderate P_{CD}	16	6	5
Subsidies High P_{CD}	5	4	3

The profitability is verified in thirteen case-studies. It ranges from 1357 €/kW (scenarios Subsidies High P_{CD} and Self-consumption 50%) to 48 €/kW (scenarios Fiscal deduction 50% and Self-consumption 30%). NPV is negative when it is hypothesized a $w_{self,c}$ equal to 30% considering or a rate of fiscal deduction of 36% or an unitary value of subsidy of 10 €/tCO$_2$eq. These values can be referred to the existing literature also when was applied a FIT scheme: 716–913 €/kW (Chiaroni et al. 2014), 1804–2386 €/kW (Campoccia et al. 2014), (−1300)–3300 €/kW (Bortolini et al. 2013).

Results proposed in this work underline that the share of self-consumption plays a role more critical than subsidies. The profitability of residential PV systems depends by this variable in a mature market (Sarasa-Maestro et al. 2016). A value of 30% is used typically in the evaluation of economic feasibility, because the production of energy from PV modules has its peak during the day, while consumers are busy to work outside the home. A possible solution to intermittent nature of this RES is represented by the application of a battery storage, but this choice requires also an appropriate environmental evaluation (Üçtuğ and Azapagic 2018). The use of intelligent machinery represents another technical solution to solve this issue (Zhou et al. 2016).

The comparison among several political tools underline as the increase of rate of fiscal deduction to 50% permits to reach better economic performance than the application of a subsidies with a low price of carbon dioxide. In addition, there is an increase of 226 €/kW applying a fiscal deduction of 50% than 36%. Consequently, the choice of subsidized fiscal deduction is useful, but the quantity of PV power installed is been low and so the market has not rewarded this choice.

The re-introduction of subsidies can have a shock effect pushing the investors to opt for this choice. In fact, starting by the idea to support the contrast to climate change when also economic opportunities are verified, the development of PV plants can involve homes in which currently renewable plants are not installed. The increase of energy self-sufficiency is a long-term objective.

NPV obtained in scenarios Subsidies Moderate P_{CD} are greater than ones of Fiscal deduction 50% and an analysis of Break-Even point notes that this point is equal to 18.50 €/tCO$_2$eq. A comparison with recent values reported in the market (see Figure 4) underlines that there is a difference very low with current values (about 1 €/tCO$_2$eq). NPV increases of 313 €/kW using a moderate P_{CD} than low P_{CD} and this increase becomes 438 €/kW when is choice a high P_{CD} than moderate P_{CD}.

The DPBT results are coherent with the NPV ones. Two unprofitable case studies are characterised by a value >20. In fact, in the worse scenario the cut-off period is fixed equal to the lifetime of the plant and when is reported a DPBT >20 the investment cannot be recovered within this interval time. The difference between DCI and DCO has always a negative sign. In three case studies (Subsidies Moderate P_{CD} with Self-consumption 30%, Fiscal deduction 50% with Self-consumption 40%

and Fiscal deduction 36% with Self-consumption 50%) has more sign changes. While, the remaining case studies have only one sign change.

DPBT varies from 3 years (scenarios Subsidies High P$_{CD}$ and Self-consumption 50%) to 19 years (scenarios Fiscal Deduction 50% and Self-consumption 30%). This result is justified by application of third-party funds that distribute the investment cost over the years of loan. Seven case studies have a value that does not exceed 6 years and it is comparable with other works: 3–12 years (Chiaroni et al. 2014), 4–8 years (Rodrigues et al. 2016) and 7–15 years (Orioli and Di Gangi 2015).

4.2. The Distribution of Revenues

The profitability is characterized by several items. An analysis of their percentage distribution can be useful to define the relevance of these variables. Obviously, the distribution depends by typology of case study—Figure 5.

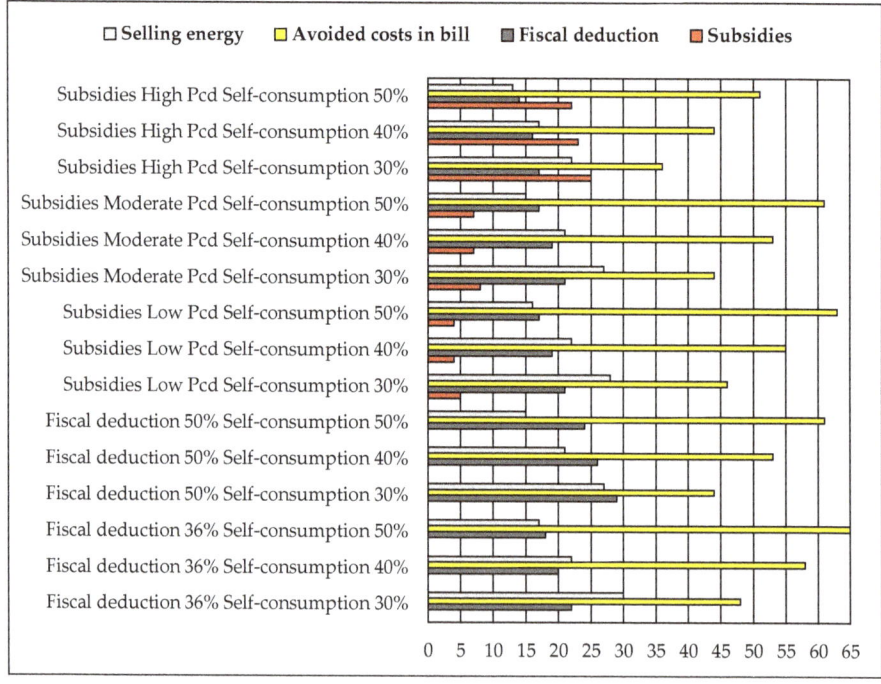

Figure 5. The distribution of revenues. Data expressed in percentage.

A consumer pays to use electricity and when a PV system is installed, the investor (consumer) becomes also a producer of energy (also called prosumer). For this motive, the purchase of energy is not more necessary (relatively to the share of self-consumption) and avoided cost of bills can be interpreted as a revenue. In all case studies, this represent the aim item of discounted cash inflow. It varies from 36% in scenarios Subsidies High P$_{CD}$ and Self-consumption 30% to 65% in scenarios Fiscal deduction 36% and Self-consumption 50%.

The selling of energy to the grid has a percentage weight basically greater than fiscal deduction. This is not verified only in scenarios in which the consumer reaches a share of self-consumption equal to 50%. The fiscal deduction permits to reduce the taxable income and it is applied following by the investment in PV system. In this way, there is a reduction of taxable costs and this item can be interpreted as a revenue.

Literature analysis underlined as subsidies played a key-role in the economic evaluation of PV plants. In this context their weight is marginal, when is hypothesized a low value of P_{CD} (about 4–5%) or a moderate value of P_{CD} (about 7–8%). Instead, they have a weight of about 22–25%, when the reduction of carbon dioxide assumes a value of 70 € per ton of CO_2.

4.3. Alternative Scenarios

NPV are obtained according to the assumptions of a set of input variables. In order to give solidity to results obtained, a sensitivity on the critical variables is conducted. In this way, a variance of the expected NPV could occur and this analysis defines the variations of this index (Sommerfeldt and Madani 2017).

Some variables are already changed in baseline case studies and for this motive the same approach is repeated in this analysis. Section 4.1 has defined as NPV varies in function of the share of self-consumption, the rate of fiscal deduction and the value of carbon dioxide (subsidies).

Alternative scenarios are constructed considering two distinct scenarios (one pessimistic and one optimistic) for the critical variables that are not examined previously (Cucchiella et al. 2017a; Sarasa-Maestro et al. 2016; Radomes and Arango 2015):

- electricity purchase price. Section 4.2 has defined that this cost, having a sign negative, can be interpreted as a revenue. The variable is decreased (Table 4) and increased (Table 5) of 0.02 cent€/kWh.
- electricity sales price. The consumer can sell to the grid the share of energy not self-consumed. The variable is decreased (Table 6) and increased (Table 7) of 0.015 cent€/kWh.
- unitary investment cost. A significant decrease of investment costs has characterised the PV market. This is caused by political choices (e.g., subsidies) that have favoured a consistent amount of installed PV systems. The variable is increased (Table 8) and decreased (Table 9) of 200 €/kW.
- average annual insolation. Italy presents several insolation levels due to its geographical conformation varying from 1350 kWh/m² × year (northern region—Table 10) to 1600 kWh/m² × year (southern region—Table 11).

Table 4. NPV in alternative scenario (p^c = 17 cent€/kWh). Data expressed in €.

Scenarios	Self-Consumption 30%	Self-Consumption 40%	Self-Consumption 50%
Fiscal Deduction 36%	−928	−72	785
Fiscal Deduction 50%	−250	606	1463
Subsidies Low P_{CD}	−553	304	1160
Subsidies Moderate P_{CD}	385	1242	2099
Subsidies High P_{CD}	1699	2555	3412

Table 5. NPV in alternative scenario (p^c = 21 cent€/kWh). Data expressed in €.

Scenarios	Self-Consumption 30%	Self-Consumption 40%	Self-Consumption 50%
Fiscal Deduction 36%	−138	982	2102
Fiscal Deduction 50%	540	1659	2779
Subsidies Low P_{CD}	237	1357	2477
Subsidies Moderate P_{CD}	1175	2295	3415
Subsidies High P_{CD}	2489	3609	4729

Table 6. NPV in alternative scenario (p^s = 4 cent€/kWh). Data expressed in €.

Scenarios	Self-Consumption 30%	Self-Consumption 40%	Self-Consumption 50%
Fiscal Deduction 36%	−989	71	1131
Fiscal Deduction 50%	−311	749	1809
Subsidies Low P_{CD}	−613	447	1507
Subsidies Moderate P_{CD}	325	1385	2445
Subsidies High P_{CD}	1638	2698	3758

Table 7. NPV in alternative scenario (p^s = 7 cent€/kWh). Data expressed in €.

Scenarios	Self-Consumption 30%	Self-Consumption 40%	Self-Consumption 50%
Fiscal Deduction 36%	−78	839	1755
Fiscal Deduction 50%	600	1516	2433
Subsidies Low P_{CD}	297	1214	2131
Subsidies Moderate P_{CD}	1235	2152	3069
Subsidies High P_{CD}	2549	3466	4382

Table 8. NPV in alternative scenario ($C_{inv,unit}$ = 2100 €/kW). Data expressed in €.

Scenarios	Self-Consumption 30%	Self-Consumption 40%	Self-Consumption 50%
Fiscal Deduction 36%	−1134	−145	843
Fiscal Deduction 50%	−385	604	1592
Subsidies Low P_{CD}	−759	230	1218
Subsidies Moderate P_{CD}	180	1168	2156
Subsidies High P_{CD}	1493	2482	3470

Table 9. NPV in alternative scenario ($C_{inv,unit}$ = 1700 €/kW). Data expressed in €.

Scenarios	Self-Consumption 30%	Self-Consumption 40%	Self-Consumption 50%
Fiscal Deduction 36%	67	1056	2044
Fiscal Deduction 50%	674	1662	2650
Subsidies Low P_{CD}	443	1431	2419
Subsidies Moderate P_{CD}	1381	2369	3357
Subsidies High P_{CD}	2694	3683	4671

Table 10. NPV in alternative scenario (t_r = 1300 kWh/m² × year). Data expressed in €.

Scenarios	Self-Consumption 30%	Self-Consumption 40%	Self-Consumption 50%
Fiscal Deduction 36%	−1094	−208	678
Fiscal Deduction 50%	−416	470	1356
Subsidies Low P_{CD}	−758	128	1014
Subsidies Moderate P_{CD}	83	970	1856
Subsidies High P_{CD}	1261	2147	3033

Table 11. NPV in alternative scenario (t_r = 1600 kWh/m² × year). Data expressed in €.

Scenarios	Self-Consumption 30%	Self-Consumption 40%	Self-Consumption 50%
Fiscal Deduction 36%	28	1118	2209
Fiscal Deduction 50%	705	1796	2886
Subsidies Low P_{CD}	442	1532	2623
Subsidies Moderate P_{CD}	1477	2568	3658
Subsidies High P_{CD}	2926	4017	5107

The profitability is verified in one-hundred and three case studies in alternative scenarios. In particular, NPV is always positive in two scenarios. The first when is applied a value of t_r equal to 1600 kWh/m^2 × year and the second is verified with a $C_{inv,unit}$ equal to 1700 €/kW. Instead, the unprofitability is obtained in seventeen case studies: fourteen when the share of self-consumption is equal to 30% (six in combination with Fiscal Deduction 36% and four with both Fiscal Deduction 50% and Subsidies Low P_{CD}) and three with a $w_{self,c}$ equal to 40% (in combination with Fiscal Deduction 36%).

This work does not assign a probability value to single case studies. However, the solar irradiation calculated in baseline scenario is subject to variation when is considered a territory situated in a northern or southern region. NPV varies from −365 €/kW to 1011 €/kW in the North of Italy, it ranges from 9 €/kW to 1702 €/kW in the South of Italy.

Italian PV market is mature and consequently, the variation of investment costs is not expected. However, the difference of costs can be proposed by several firms in order to expand their market share. NPV ranges from −378 €/kW to 1157 €/kW when is considered an increase of costs in comparison to baseline scenario, while it varies from 22 €/kW to 1557 €/kW in the opposite situation.

Regarding electricity sales price, a possible variation can be assumed when is applied a Net Metering Scheme, in which the price of electricity is increased above market value. The development of decentralized energy systems aims to obtain that single units must be self-sufficient in terms of energy and consequently, all advantages must be destined to the share of self-consumption. NPV varies from −330 €/kW to 1253 €/kW with a p^s equal to 4 cent€/kWh and it ranges from −26 €/kW to 1461 €/kW with a p^s equal to 7 cent€/kWh.

The energy bill is composed by several components and its value depends by time bands. Currently, there is an increase of energy bill in Italy. For this motive, there is a concrete opportunity that scenario presented in Table 5 can be real. NPV varies from −46 €/kW to 1576 €/kW. While, in the opposition situation (p^c equal to 17 cent€/kWh) it ranges from −309 €/kW to 1137 €/kW.

Finally, alternative values RECD can of be analysed. In Section 3.1, energetic mix is calculated at net of renewables and imports and RECD is equal to 685 gCO$_2$eq/kWh. However, PV plant can be compared with an energy portfolio in which also renewables and imports are considered. Initially, the distribution of energy sources is evaluated for 2017 (ENEA 2018): PEM_{GAS} = 36.5%, PEM_{OIL} = 34%, PEM_{COAL} = 6%, PEM_{RES} = 19% (percentage in energy mix of renewables) and PEM_{IMP} = 4.5% (percentage in energy mix of imports). In particular, renewables can be subdivided in hydroelectric (HYD), PV, biomass (BIO), wind (WIN) and geothermal (GEO). Their distribution is calculated according to values of GSE (Gestore Servizi Energetici) regarding electricity sector in 2016: PEM_{HYD} = 39%, PEM_{PV} = 21%, PEM_{BIO} = 18%, PEM_{WIN} = 16% and PEM_{GEO} = 6%. The values of emissions are chosen as average values of Figure 3: ECD_{HYD} = 19 gCO$_2$eq/kWh, ECD_{PV} = 42 gCO$_2$eq/kWh, ECD_{BIO} = 152 gCO$_2$eq/kWh, ECD_{WIN} = 34 gCO$_2$eq/kWh and ECD_{GEO} = 41 gCO$_2$eq/kWh. For the value of imports is considered the average among oil, carbon and gas (ECD_{IMP} = 847 gCO$_2$eq/kWh)—Equation (16).

$$RECD = (ECD_{OIL} \times PEM_{OIL} + ECD_{COAL} \times PEM_{COAL} + ECD_{GAS} \times PEM_{GAS} + ECD_{HYD} \times PEM_{HYD} + ECD_{PV} \times PEM_{PV} + ECD_{WIN} \times PEM_{WIN} + ECD_{BIO} \times PEM_{BIO} + ECD_{GEO} \times PEM_{GEO} + ECD_{IMP} \times PEM_{IMP}) - ECD_{PV} = 604 - 42 = 562 \text{ gCO}_2\text{eq/kWh} \quad (16)$$

According to Equations (13)–(15), the following step is the transformation of environmental benefits in economic terms. The unitary value of subsidies changes in function of the value of P_{CD}—Equations (17)–(19).

$$SUB_{PV,1} = 562 \times 10 \times 4680 = 26 \text{ €/year} \quad \text{Subsidies Low } P_{CD} \quad (17)$$

$$SUB_{PV,1} = 562 \times 35 \times 4680 = 92 \text{ €/year} \quad \text{Subsidies Moderate } P_{CD} \quad (18)$$

$$\text{SUB}_{PV,1} = 562 \times 70 \times 4680 = 184\,\text{€/year} \quad \text{Subsidies High P}_{CD} \tag{19}$$

The variation of NPV in alternative scenarios in which RECD is assumed equal to 562 gCO₂eq/kWh is proposed in Table 12. Obviously, both scenarios Fiscal Deduction 36% and Fiscal Deduction 50% are not modified by this change.

Table 12. NPV in alternative scenario (RECD = 562 gCO_2eq/kWh). Data expressed in €.

Scenarios	Self-Consumption 30%	Self-Consumption 40%	Self-Consumption 50%
Subsidies Low P_{CD}	−225	763	1751
Subsidies Moderate P_{CD}	544	1533	2521
Subsidies High P_{CD}	1622	2610	3599

The profitability is confirmed in several scenarios (only scenario Subsidies Low P_{CD} and Self-consumption 30% has a negative NPV). The index varies from 181 €/kW to 1200 €/kW. The presence of renewable in an energy mix determines a reduction of carbon dioxide linked to this portfolio. In fact, a reduction of RECD is verified and it is equal to 123 gCO₂eq/kWh. This determines a reduction of value of SUB$_{PV}$ and consequently, also NPV is characterized by a reduction. It varies from about 20 €/kW (Self-consumption 30%) to 160 €/kW (Self-consumption 50%).

4.4. Discussions and Policy Implications

The transition towards a low carbon society requires to evaluate the relationship between the CE models and the use of REs. The CE framework is characterized by requirements to measure. One of them is increasing share of renewable and recyclable resources.

The reduction of GHG emissions is possible thanks to the use of less raw materials and more sustainable sourcing (Elia et al. 2017). Recycling and recovery of materials as indium, silicon and silver can be obtained by PV waste favouring the application of CE model (Brenner and Adamovic 2017). However, a sustainable RE technology requires that the all parts of the product lifecycle can be optimized. The analysis from cradle to growth is conducted (Charles et al. 2016). The recovery of PV modules is typically characterized by unprofitability (Choi and Fthenakis 2014).

CE model aims to favour the development of REs and economic opportunities take the front seat (Kopnina 2017). This work follows this approach. In fact, PV investment is characterized by a low risk and results obtained define that the profitability can reach interesting values.

PV systems are able not only to favour the decarbonisation of society, but also to reduce geopolitical risks. In fact, when a country increases the internally energy produced there is also a decrease in external energy required. Consumers can increase their profits in a significant way and this is possible through the harmonization between demanded and produced energy. At the same time, consumers are responsible actors towards targets to reach.

Energy firms move from centralised to decentralised power and new business models emerge in which people provide the energy for their homes and commercial premises. At the same time, emissions constraints for manufacturing of products represent another motivation to develop REs.

Subsidies cannot be seen as a perpetual assistance, but in this new proposal consumers sell the amount of CO₂eq avoided using a PV system instead to use electricity by fossil fuels.

The Paris Agreement is a crucial step to reduce the decarbonisation of society. A mix of renewable resources, energy efficiency, an appropriate waste management and material efficiency strategies represent initiatives to implement. In this way, renewable economy and circular economy moves towards the same direction.

5. Conclusions

Renewables represent the main actor in a transition towards a society low-carbon. PV source plays a key-role, in fact its growth has assumed significant values in the last years globally. However, consumers are also investors and a project is implemented only if economic conditions are verified.

Currently, a subsidized fiscal rate of 50% (instead of 36%) is applied to the Italian context. This measure has not produced a consistent increase of the power installed. A new proposal is defined in this work, in which when a consumer reduces carbon dioxide emission levels has right to receive an economic contribution. This is paid by operators that emit a level of pollutants greater than the value allowed (carbon price defined by a real market). In addition, this incentive is given to the energy produced by a PV plant for all its lifetime (20 years). The subsidy is assumed fixed according to the FIT scheme.

Literature review has underlined as a consistent quantity of emissions is not covered by a carbon price and this value is below 10 €/tCO_2eq. However, several authors have highlighted as carbon price must have a greater value in order to tackle climate change.

The reduction of carbon dioxide is calculated according to values reported in literature. Italy moves towards a reduction of use of both oil and carbon, at the same time there is an increase of natural gas. The share of RE tends to be stable. The reduction of emissions is assumed equal to 685 gCO_2eq/kWh and the market value of EU ETS is characterised by an increase of about 13 €/tCO_2eq considering July 2017–July 2018 as interval period.

The fiscal deduction with a rate of 50% produces more profits for consumers in comparison to a subsidy determined by the price of CO_2eq when this value is lower than 18.50 €/tCO_2eq. Among this value and one proposed by market there is a difference of only 1 €/tCO_2eq. Consequently, this choice can be applied in a real context.

This analysis follows the values reported by report of the High-Level Commission on Carbon Prices. Applying a price of carbon dioxide equal to 35 €/tCO_2eq. NPV varies from 260 €/kW to 919 €/kW and DPBT can be equal to 5–6 years. When, instead, is applied a price of carbon dioxide equal to 70 €/tCO_2eq, NPV ranges from 698 €/kW to 1357 €/kW and DPBT varies from 3 years to 6 years.

Profits obtained are probably not relevant, but consumer can opt towards this choice for the following aspects: (i) investment costs are low, (ii) reduces the costs of energy bill and can also obtained profits and (iii) contributes to tackle the climate change. The harmonization between demanded and produced energy increases the economic performance. Alternative scenarios give solidity to results obtained.

A new development of residential PV applications is able to increase the sustainability of a country and the quantitative analysis proposed in this work demonstrates as PV source contributes to the CE models.

Funding: This research received no external funding.

Conflicts of Interest: The author declares no conflict of interest.

References

Aatola, Piia, Markku Ollikainen, and Anne Toppinen. 2013. Price Determination in the Eu Ets Market: Theory and Econometric Analysis with Market Fundamentals. *Energy Economics* 36: 380–95. [CrossRef]

Alberici, Sacha, Sil Boeve, Pieter van Breevoort, Yvonne Deng, Sonja Förster, Ann Gardiner, Valentijn van Gastel, Katharina Grave, Heleen Groenenberg, David de Jager, and et al. 2014. Subsidies and Costs of Eu Energy—Final Report and Annex 3, Ecofys. Available online: https://ec.europa.eu/energy/en/content/final-report-ecofys (accessed on 9 July 2018).

Alberola, Emilie, Julien Chevallier, and Benoît Chèze. 2008. Price Drivers and Structural Breaks in European Carbon Prices 2005–2007. *Energy Policy* 36: 787–97. [CrossRef]

Avril, Sophie, Christine Mansilla, Marie Busson, and Thibault Lemaire. 2012. Photovoltaic Energy Policy: Financial Estimation and Performance Comparison of the Public Support in Five Representative Countries. *Energy Policy* 51: 244–58. [CrossRef]

Bakhtyar, Bardia, Ahmad Fudholi, Kabir Hassan, Muhammad Azam, Chin Haw Lim, Ngai Weng Chan, and Kamaruzzaman Sopian. 2017. Review of CO_2 Price in Europe Using Feed-in Tariff Rates. *Renewable and Sustainable Energy Reviews* 69: 685–91. [CrossRef]

Baur, Lucia, and Mauricio Uriona. 2018. Diffusion of Photovoltaic Technology in Germany: A Sustainable Success or an Illusion Driven by Guaranteed Feed-in Tariffs? *Energy* 150: 289–98. [CrossRef]

Böhringer, Christoph, Bouwe Dijkstra, and Knut Einar Rosendahl. 2014. Sectoral and Regional Expansion of Emissions Trading. *Resource and Energy Economics* 37: 201–25. [CrossRef]

Bortolini, Marco, Mauro Gamberi, Alessandro Graziani, Cristina Mora, and Alberto Regattieri. 2013. Multi-Parameter Analysis for the Technical and Economic Assessment of Photovoltaic Systems in the Main European Union Countries. *Energy Conversion and Management* 74: 117–28. [CrossRef]

Bravi, Mirko, Maria Laura Parisi, Enzo Tiezzi, and Riccardo Basosi. 2011. Life Cycle Assessment of a Micromorph Photovoltaic System. *Energy* 36: 4297–306. [CrossRef]

Brenner, W., and Nadja Adamovic. 2017. A Circular Economy for Photovoltaic Waste-the Vision of the European Project Cabriss. Paper presented at the 40th International Convention on Information and Communication Technology, Electronics and Microelectronics, Opatija, Croatia, May 22–26.

Breyer, Christian, Dmitrii Bogdanov, Ashish Gulagi, Arman Aghahosseini, Larissa S. N. S. Barbosa, Otto Koskinen, Maulidi Barasa, Upeksha Caldera, Svetlana Afanasyeva, and Michael Child. 2017. On the Role of Solar Photovoltaics in Global Energy Transition Scenarios. *Progress in Photovoltaics: Research and Applications* 25: 727–45. [CrossRef]

Cai, Wugan, and Jiafeng Pan. 2017. Stochastic Differential Equation Models for the Price of European CO_2 Emissions Allowances. *Sustainability* 9: 207. [CrossRef]

Campoccia, Angelo, Luigi Dusonchet, Enrico Telaretti, and Gaetano Zizzo. 2014. An Analysis of Feed'in Tariffs for Solar Pv in Six Representative Countries of the European Union. *Solar Energy* 107: 530–42. [CrossRef]

Charles, Rhys G., Matthew L. Davies, and Peter Douglas. 2016. Third Generation Photovoltaics—Early Intervention for Circular Economy and a Sustainable Future. Paper presented at the 2016 Electronics Goes Green, Berlin, Germany, September 7–9.

Chiaroni, Davide, Vittorio Chiesa, Lorenzo Colasanti, Federica Cucchiella, Idiano D'Adamo, and Federico Frattini. 2014. Evaluating Solar Energy Profitability: A Focus on the Role of Self-Consumption. *Energy Conversion and Management* 88: 317–31. [CrossRef]

Choi, Jun-Ki, and Vasilis Fthenakis. 2014. Crystalline Silicon Photovoltaic Recycling Planning: Macro and Micro Perspectives. *Journal of Cleaner Production* 66: 443–49. [CrossRef]

Christiansen, Atle C., Andreas Arvanitakis, Kristian Tangen, and Henrik Hasselknippe. 2005. Price Determinants in the Eu Emissions Trading Scheme. *Climate Policy* 5: 15–30. [CrossRef]

Comello, Stephen, and Stefan Reichelstein. 2017. Cost Competitiveness of Residential Solar Pv: The Impact of Net Metering Restrictions. *Renewable and Sustainable Energy Reviews* 75: 46–57. [CrossRef]

Convery, Frank J., and Luke Redmond. 2007. Market and Price Developments in the European Union Emissions Trading Scheme. *Review of Environmental Economics and Policy* 1: 88–111. [CrossRef]

Cucchiella, Federica, Idiano D'Adamo, and Massimo Gastaldi. 2016. A Profitability Assessment of Small-Scale Photovoltaic Systems in an Electricity Market without Subsidies. *Energy Conversion and Management* 129: 62–74. [CrossRef]

Cucchiella, Federica, Idiano D'Adamo, and Massimo Gastaldi. 2017a. Economic Analysis of a Photovoltaic System: A Resource for Residential Households. *Energies* 10: 814. [CrossRef]

Cucchiella, Federica, Idiano D'Adamo, and Massimo Gastaldi. 2017b. The Economic Feasibility of Residential Energy Storage Combined with Pv Panels: The Role of Subsidies in Italy. *Energies* 10: 1434. [CrossRef]

Dadural, Joshua S., and Leah R. Reznikov. 2018. Interest in and Awareness of French President Emmanuel Macron's "Make Our Planet Great Again" Initiative. *Social Sciences* 7: 102. [CrossRef]

Dirix, Jo, Wouter Peeters, and Sigrid Sterckx. 2015. Is the Eu Ets a Just Climate Policy? *New Political Economy* 20: 702–24. [CrossRef]

Doytch, Nadia, and Merih Uctum. 2016. Globalization and the Environmental Impact of Sectoral Fdi. *Economic Systems* 40: 582–94. [CrossRef]

Edenhofer, Ottmar, Ramón Pichs-Madruga, Youba Sokona, Christopher Field, Vicente Barros, Thomas Stocker, Qin Dahe, Jan Minx, Katharine Mach, and Gian-Kasper Plattner. 2012. Renewable Energy Sources and Climate Change Mitigation—Special Report of the Intergovernmental Panel on Climate Change. Available online: http://Www.Ipcc.Ch/ (accessed on 9 July 2018).

Elia, Valerio, Maria Grazia Gnoni, and Fabiana Tornese. 2017. Measuring Circular Economy Strategies through Index Methods: A Critical Analysis. *Journal of Cleaner Production* 142: 2741–51. [CrossRef]

ENEA. 2018. Analisi Trimestrale Del Sistema Energetico Italiano—Anno 2017. Available online: http://www.Enea.It/It (accessed on 9 July 2018).

Escoto Castillo, Ana, and Landy Sánchez Peña. 2017. Diffusion of Electricity Consumption Practices in Mexico. *Social Sciences* 6: 144. [CrossRef]

European Commision. 2016. The Eu Emissions Trading System (Eu Ets). Available online: https://ec.europa.eu/clima/policies/ets_en (accessed on 28 June 2018).

Fang, Guochang, Lixin Tian, Menghe Liu, Min Fu, and Mei Sun. 2018a. How to Optimize the Development of Carbon Trading in China—Enlightenment from Evolution Rules of the Eu Carbon Price. *Applied Energy* 211: 1039–49. [CrossRef]

Fang, Guochang, Lixin Tian, Min Fu, Mei Sun, Yu He, and Longxi Lu. 2018b. How to Promote the Development of Energy-Saving and Emission-Reduction with Changing Economic Growth Rate—A Case Study of China. *Energy* 143: 732–45. [CrossRef]

Fthenakis, Vasilis M., and Hyung Chul Kim. 2013. Life Cycle Assessment of High-Concentration Photovoltaic Systems. *Progress in Photovoltaics: Research and Applications* 21: 379–88. [CrossRef]

Fthenakis, Vasilis M., Hyung Chul Kim, and Erik Alsema. 2008. Emissions from Photovoltaic Life Cycles. *Environmental Science & Technology* 42: 2168–74.

Gerbeti, Agime. 2016. CO_2 in Goods. In *Green Fiscal Reform for a Sustainable Future: Reform, Innovation and Renewable Energy*. Cheltenham: Edward Elgar Publishing, pp. 77–92.

Gerbeti, Agime. 2017. Mercati Ambientali. Aggiornamento 2017. *Enciclopedia Treccani* 2: 107–11.

Gerbeti, Agime. 2018. Sustainability as a Parameter for Industrial Competitivenes. *Energia Elettrica* 95: 13–21.

Hosenuzzaman, Md, Nasrudin Abd Rahim, Jeyraj A. L. Selvaraj, Mohammad Hasanuzzaman, A. B. M. Abdul Malek, and Afroza Nahar. 2015. Global Prospects, Progress, Policies, and Environmental Impact of Solar Photovoltaic Power Generation. *Renewable and Sustainable Energy Reviews* 41: 284–97. [CrossRef]

Hou, Guofu, Honghang Sun, Ziying Jiang, Ziqiang Pan, Yibo Wang, Xiaodan Zhang, Ying Zhao, and Qiang Yao. 2016. Life Cycle Assessment of Grid-Connected Photovoltaic Power Generation from Crystalline Silicon Solar Modules in China. *Applied Energy* 164: 882–90. [CrossRef]

IEA. 2015. Global Energy & CO_2 Status Report. Available online: https://Webstore.Iea.Org/ (accessed on 28 June 2018).

Isacs, Lina, Göran Finnveden, Lisbeth Dahllöf, Cecilia Håkansson, Linnea Petersson, Bengt Steen, Lennart Swanström, and Anna Wikström. 2016. Choosing a Monetary Value of Greenhouse Gases in Assessment Tools: A comprehensive Review. *Journal of Cleaner Production* 127: 37–48. [CrossRef]

Keppler, Jan Horst, and Maria Mansanet-Bataller. 2010. Causalities between CO_2, Electricity, and Other Energy Variables during Phase I and Phase II of the Eu Ets. *Energy Policy* 38: 3329–41. [CrossRef]

Khan, Jibran, and Mudassar H. Arsalan. 2016. Solar Power Technologies for Sustainable Electricity Generation—A Review. *Renewable and Sustainable Energy Reviews* 55: 414–25. [CrossRef]

Kopnina, Helen. 2017. European Renewable Energy. Applying Circular Economy Thinking to Policy-Making. *Visions for Sustainability* 8: 7–19.

Landis, Florian, and Thomas Bernauer. 2012. Transfer Payments in Global Climate Policy. *Nature Climate Change* 2: 628. [CrossRef]

Lee, Minhyun, Taehoon Hong, Choongwan Koo, and Chan-Joong Kim. 2017. A Break-Even Analysis and Impact Analysis of Residential Solar Photovoltaic Systems Considering State Solar Incentives. *Technological and Economic Development of Economy* 24: 358–82. [CrossRef]

Lin, Shu-Kun. 2012. Social Sciences and Sustainability. *Social Sciences* 1: 1. [CrossRef]

Lin, Boqiang, and Yufang Chen. 2018. Carbon Price in China: A CO_2 Abatement Cost of Wind Power Perspective. *Emerging Markets Finance and Trade* 54: 1653–71. [CrossRef]

Louwen, Atse, Wilfried G. J. H. M. van Sark, André P. C. Faaij, and Ruud E. I. Schropp. 2016. Re-Assessment of Net Energy Production and Greenhouse Gas Emissions Avoidance after 40 Years of Photovoltaics Development. *Nature Communications* 7: 13728. [CrossRef] [PubMed]

Markets Insider. 2018. CO_2 European Emission Allowances in Eur—Historical Prices. Available online: https://Markets.Businessinsider.Com/Commodities/Historical-Prices/CO2-Emissionsrechte/Euro/26.6.2017_29.7.2018 (accessed on 29 July 2018).

Mauleón, Ignacio. 2017. Photovoltaic Investment Roadmaps and Sustainable Development. *Journal of Cleaner Production* 167: 1112–21. [CrossRef]

Metivier, Clement, Sebastien Postic, Emilie Alberola, and Madhulika Vinnakota. 2017. Global Panorama of Carbon Prices in 2017. Available online: https://www.I4ce.Org/ (accessed on 28 June 2018).

OECD. 2016. Effective Carbon Rates on Energy. Available online: http://www.Oecd.Org/ (accessed on 9 July 2018).

Ogland-Hand, Jonathan D., Jeffrey M. Bielicki, and Thomas A. Buscheck. 2017. The Value of CO_2-Bulk Energy Storage to Reducing CO_2 Emissions. *Energy Procedia* 114: 6886–92. [CrossRef]

Olson, Arne, Chi-Keung Woo, Nick Schlag, and Alison Ong. 2016. What Happens in California Does Not Always Stay in California: The Effect of California's Cap-and-Trade Program on Wholesale Electricity Prices in the Western Interconnection. *The Electricity Journal* 29: 18–22. [CrossRef]

Orioli, Aldo, and Alessandra Di Gangi. 2015. The Recent Change in the Italian Policies for Photovoltaics: Effects on the Payback Period and Levelized Cost of Electricity of Grid-Connected Photovoltaic Systems Installed in Urban Contexts. *Energy* 93 Part 2: 1989–2005. [CrossRef]

Orioli, Aldo, Vincenzo Franzitta, Alessandra Di Gangi, and Ferdinando Foresta. 2016. The Recent Change in the Italian Policies for Photovoltaics: Effects on the Energy Demand Coverage of Grid-Connected Pv Systems Installed in Urban Contexts. *Energies* 9: 944. [CrossRef]

Park, Hojeong, and Won Kyung Hong. 2014. Korea's Emission Trading Scheme and Policy Design Issues to Achieve Market-Efficiency and Abatement Targets. *Energy Policy* 75: 73–83. [CrossRef]

Parry, Ian, Chandara Veung, and Dirk Heine. 2015. How Much Carbon Pricing Is in Countries' Own Interests? The Critical Role of Co-Benefits. *Climate Change Economics* 6: 1550019. [CrossRef]

Peng, Jinqing, Lin Lu, and Hongxing Yang. 2013. Review on Life Cycle Assessment of Energy Payback and Greenhouse Gas Emission of Solar Photovoltaic Systems. *Renewable and Sustainable Energy Reviews* 19: 255–74. [CrossRef]

Pyrgou, Andri, Angeliki Kylili, and Paris A. Fokaides. 2016. The Future of the Feed-in Tariff (Fit) Scheme in Europe: The Case of Photovoltaics. *Energy Policy* 95: 94–102. [CrossRef]

Radomes, Amando A., Jr., and Santiago Arango. 2015. Renewable Energy Technology Diffusion: An Analysis of Photovoltaic-System Support Schemes in Medellín, Colombia. *Journal of Cleaner Production* 92: 152–61. [CrossRef]

REN21. 2018. Renewables 2018 Global Status Report. Available online: http://Www.Ren21.Net/Gsr-2017/ (accessed on 28 June 2018).

Rodrigues, Sandy, Roham Torabikalaki, Fábio Faria, Nuno Cafôfo, Xiaoju Chen, Ashkan Ramezani Ivaki, Herlander Mata-Lima, and F. Morgado-Dias. 2016. Economic Feasibility Analysis of Small Scale Pv Systems in Different Countries. *Solar Energy* 131: 81–95. [CrossRef]

Saavedra, Marroquin, Melkyn Ricardo, Cristiano Hora de O. Fontes, and Francisco Gaudêncio M. Freires. 2018. Sustainable and Renewable Energy Supply Chain: A System Dynamics Overview. *Renewable and Sustainable Energy Reviews* 82: 247–59. [CrossRef]

Sampaio, Priscila Gonçalves Vasconcelos, and Mario Orestes Aguirre González. 2017. Photovoltaic Solar Energy: Conceptual Framework. *Renewable and Sustainable Energy Reviews* 74: 590–601. [CrossRef]

Sarasa-Maestro, Carlos, Rodolfo Dufo-López, and José Bernal-Agustín. 2016. Analysis of Photovoltaic Self-Consumption Systems. *Energies* 9: 681. [CrossRef]

Segura, Sara, Luis Ferruz, Pilar Gargallo, and Manuel Salvador. 2018. Environmental Versus Economic Performance in the Eu Ets from the Point of View of Policy Makers: A Statistical Analysis Based on Copulas. *Journal of Cleaner Production* 176: 1111–32. [CrossRef]

Smith, Stephen, and Nils Axel Braathen. 2015. Monetary Carbon Values in Policy Appraisal. Available online: https://www.Oecd-Ilibrary.Org/Environment/Monetary-Carbon-Values-in-Policy-Appraisal_5jrs8st3ngvh-En (accessed on 9 July 2018).

Sommerfeldt, Nelson, and Hatef Madani. 2017. Revisiting the Techno-Economic Analysis Process for Building-Mounted, Grid-Connected Solar Photovoltaic Systems: Part Two—Application. *Renewable and Sustainable Energy Reviews* 74: 1394–404. [CrossRef]

Stiglitz, Joseph E., Nicholas Stern, Maosheng Duan, Ottmar Edenhofer, Gaël Giraud, Geoff Heal, Emilio Lebre la Rovere, Adele Morris, Elisabeth Moyer, Mari Pangestu, and et al. 2017. Report of the High-Level Commission on Carbon Prices Supported—Carbon Pricing Leadership Coalition. Available online: https://www.Carbonpricingleadership.Org/Report-of-the-Highlevel-Commission-on-Carbon-Prices/ (accessed on 9 July 2018).

Strupeit, Lars, and Alvar Palm. 2016. Overcoming Barriers to Renewable Energy Diffusion: Business Models for Customer-Sited Solar Photovoltaics in Japan, Germany and the United States. *Journal of Cleaner Production* 123: 124–36. [CrossRef]

Sun, Jingqi, Jing Shi, Boyang Shen, Shuqing Li, and Yuwei Wang. 2018. Nexus among Energy Consumption, Economic Growth, Urbanization and Carbon Emissions: Heterogeneous Panel Evidence Considering China's Regional Differences. *Sustainability* 10: 1–16. [CrossRef]

Tanaka, Yugo, Andrew Chapman, Shigeki Sakurai, and Tetsuo Tezuka. 2017. Feed-in Tariff Pricing and Social Burden in Japan: Evaluating International Learning through a Policy Transfer Approach. *Social Sciences* 6: 127. [CrossRef]

Tian, Lixin, Jianglai Pan, Ruijin Du, Wenchao Li, Zaili Zhen, and Gao Qibing. 2017. The Valuation of Photovoltaic Power Generation under Carbon Market Linkage Based on Real Options. *Applied Energy* 201: 354–62. [CrossRef]

Üçtuğ, Fehmi Görkem, and Adisa Azapagic. 2018. Environmental Impacts of Small-Scale Hybrid Energy Systems: Coupling Solar Photovoltaics and Lithium-Ion Batteries. *Science of The Total Environment* 643: 1579–89. [CrossRef]

Vlachou, Andriana, and Georgios Pantelias. 2017. The Eu's Emissions Trading System, Part 2: A Political Economy Critique. *Capitalism Nature Socialism* 28: 108–27. [CrossRef]

Xiong, Ling, Bo Shen, Shaozhou Qi, Lynn Price, and Bin Ye. 2017. The Allowance Mechanism of China's Carbon Trading Pilots: A Comparative Analysis with Schemes in Eu and California. *Applied Energy* 185: 1849–59. [CrossRef]

Yang, Baochen, Chuanze Liu, Yunpeng Su, and Xin Jing. 2017. The Allocation of Carbon Intensity Reduction Target by 2020 among Industrial Sectors in China. *Sustainability* 9: 148. [CrossRef]

Yang, Baochen, Chuanze Liu, Zehao Gou, Jiacheng Man, and Yunpeng Su. 2018. How Will Policies of China's CO_2 Ets Affect Its Carbon Price: Evidence from Chinese Pilot Regions. *Sustainability* 10: 605. [CrossRef]

Zhao, Xin, Meng Han, Lili Ding, and Wanglin Kang. 2018. Usefulness of Economic and Energy Data at Different Frequencies for Carbon Price Forecasting in the Eu Ets. *Applied Energy* 216: 132–41. [CrossRef]

Zhou, Nan, Nian Liu, Jianhua Zhang, and Jinyong Lei. 2016. Multi-Objective Optimal Sizing for Battery Storage of Pv-Based Microgrid with Demand Response. *Energies* 9: 591. [CrossRef]

© 2018 by the author. Licensee MDPI, Basel, Switzerland. This article is an open access article distributed under the terms and conditions of the Creative Commons Attribution (CC BY) license (http://creativecommons.org/licenses/by/4.0/).

Article

Adopting Circular Economy at the European Union Level and Its Impact on Economic Growth

Mihail Busu

Faculty of Business Administration in Foreign Languages, The Bucharest University of Economics Studies, 010374 Bucharest, Romania; mihail.busu@fabiz.ase.ro

Received: 12 April 2019; Accepted: 17 May 2019; Published: 24 May 2019

Abstract: Based on the findings of the economic studies on the implications of industrialization in the case of growing economies, this study aims to present the economic factors that are at the basis of the development of circular economy at the European Union level. Starting with the model of economic growth based on the recycling rate of municipal waste, human capital, productivity of the resources, and green energy use, three statistical hypotheses were validated through a panel data model with the use of EViews 10 statistical software. The analysis was conducted for 27 European Union countries during the time frame 2008–2017. The paper highlights that the circular economy model is determined by resource productivity, labor employed in environmental protection, recycling rate of municipal waste, and renewable energy use.

Keywords: circular economy; environmental assessment; quantitative analysis; waste management; renewable energy; economic growth; sustainable development

1. Introduction

The use of limited resources generates multiple concerns for government, as well as for academics, as they seek to find the best solution to the challenge of growing demands for consumerist economies and climate change. Ratifying the Kyoto Protocol and enforcing it in as many countries as possible creates the premise for improving the industrial processes that generate pollution.

At present, the real problem is how to change the current structure of the consumption pattern, based on a production–consumption–waste model, into a circular economy (CE) which is regenerative by definition, based on a production–consumption–reuse model. According to Kirchherr et al. (2017), a CE is most frequently depicted as a combination of reduce, reuse, and recycle activities. Thus, an essential role in the circular economy is to invest in innovative equipment for environmental protection (Porter and Van der Linde 1995).

The economic literature of the past decades abounds with econometric and economic studies quantifying the impact of environment management programs and waste on the economic development of countries, with regard to a general equilibrium model.

In order to find the right answer to the Swedish Parliament's request for a percentage reduction in the amount of waste related to the growth indicator, Sjöström and Östblom (2009) analyzed the interconnection of solid waste program management in the context of a general equilibrium model. McDonough and Braungart (2002) introduced the Cradle to Cradle (C2C) concept, which involves the recycling of waste and its transformation to new products. It is used in economic analyses which evaluate the production of renewable and clean energy, the diversity of ecosystems, and the use of the green energy sources (Browne et al. 2009). In contrast to this, other authors (Ayres 1995; De Wolf et al. 2017) have criticized the C2C formula used in industrial processes where products are transformed to waste, usually without being reused.

Beyond the evolution of the classical model of economic growth, the authors conceptualize the economic growth model to determine the main factors of impact, then capitalize empirical data in an attempt to determine the economic factors that stimulate or inhibit the transition to a circular pattern.

This paper is structured as follows: Firstly, we present an analysis of the key indicators of the CE at the EU level. Then, a description of the multiple linear regression model is discussed. Finally, the research hypotheses are presented and tested.

2. Materials and Methods

2.1. Research Methodology

The relationship between circular economy and economic growth has been analyzed by many researchers. It was demonstrated that there is a close link between the use of circular economy and economic growth (Browne et al. 2009). Other authors (Grossman and Krueger 1995; Brock and Taylor 2005; Lyasnikov et al. 2014) concluded that human capital and innovation for environmental benefits have a positive impact on economic growth.

Moreover, while some researchers (Su et al. 2013; Gopal et al. 2013; Cappa et al. 2016) argue that the use of renewable energy has a strong impact on economic growth, other economists (Cotae 2015; Ghisellini et al. 2016; Clodnitchi and Cristian 2017) conclude that the innovative enterprises that bring innovative new products with environmental benefits to the market have a bigger impact on economic growth. Nevertheless, Geng Yong et al. (2012) and George et al. (2015) argued that the productivity of resources and the recycling rate have a direct and significant impact on economic growth.

Starting from the empirical studies mentioned above, we focused our study on the research question: "What is the impact of the implementation of the circular economy on the EU's economic growth?" In addition to what is known in this area, we tried to estimate which of five independent factors (i.e., labor productivity, labor force engaged in the production of environmental goods, recycling rate of municipal waste, the share of innovative enterprises that have brought innovative new products to the market with environmental benefits, and the use of renewable energy) have the most significant impact on the dependent variable of the regression model. According to the studies mentioned above, these exogenous variables are some of the main important factors to describe circular economy.

In order to quantify this impact analysis, three statistical assumptions were formulated, as shown in Table 1.

Table 1. Hypotheses of the research study.

Hypothesis 1	European Union (EU) member states with a higher number of employees in the field of the production of goods for environmental protection have higher economic growth.
Hypothesis 2	Renewable energy use at the EU level has a significant and strong impact on economic growth.
Hypothesis 3	Innovative enterprises with big market shares in the EU member states which have brought innovative new products with environmental benefits to the market have a greater impact on economic growth.

These statistical hypotheses were tested with a multilinear time series regression model, which is described in Section 3.

2.2. Description of the Circular Economy at the EU Level

In contrast to a linear economy, a CE is based on an economic model that offers resources better value and use. In this study, a few economic indicators describing a CE with direct impact on economic growth were used, and they were proxy variables in the regression model used in the next chapter.

Figures 1–4 give us an overview of the degree of development and use of the CE at the EU level.

An important indicator of the CE is the "resource productivity" (Blomsma and Brennan 2017). This is defined as the ratio of a country's GDP to the domestic consumption of materials, and shows us the economy's efficiency in the 27 EU member states to use materials to produce well-being (Haas et al. 2015). Figure 1 shows the value of this indicator, calculated in euro/kg, at the level of EU member states.

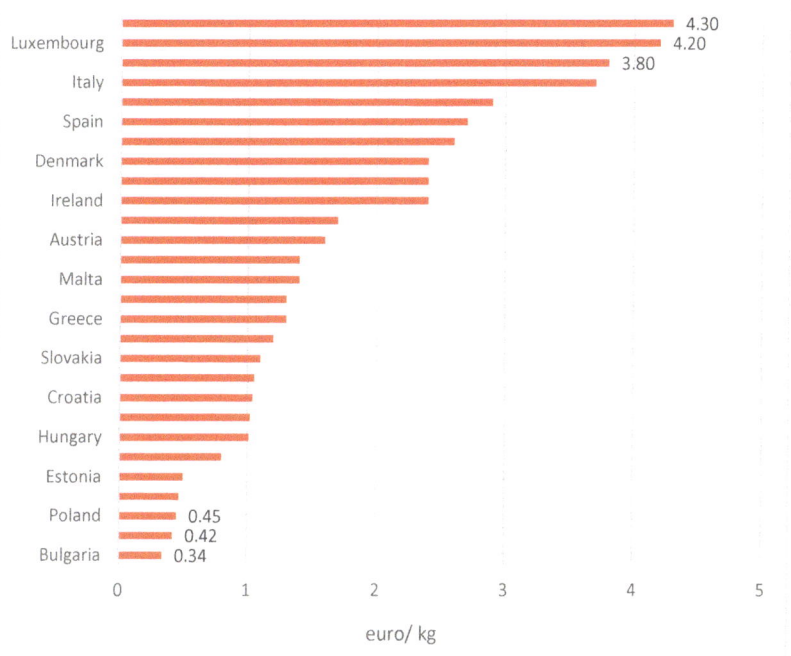

Figure 1. Productivity of the resources in EU countries in 2017. Source: based on processed data provided by Eurostat (2017).

This chart shows that the most efficient EU member states in terms of material use were the Netherlands (EUR 4.30/kg), Luxembourg (4.20 EUR/kg), and the UK (3.80 EUR/kg), while the least efficient were Poland (EUR 0.45/kg), Romania (EUR 0.42/kg), and Bulgaria (EUR 0.34/kg).

Another important indicator of the CE is given by the workforce engaged in the production of environmental goods (Lundvall 1996). This indicator can be seen at the level of the EU member states in 2017 in Figure 2.

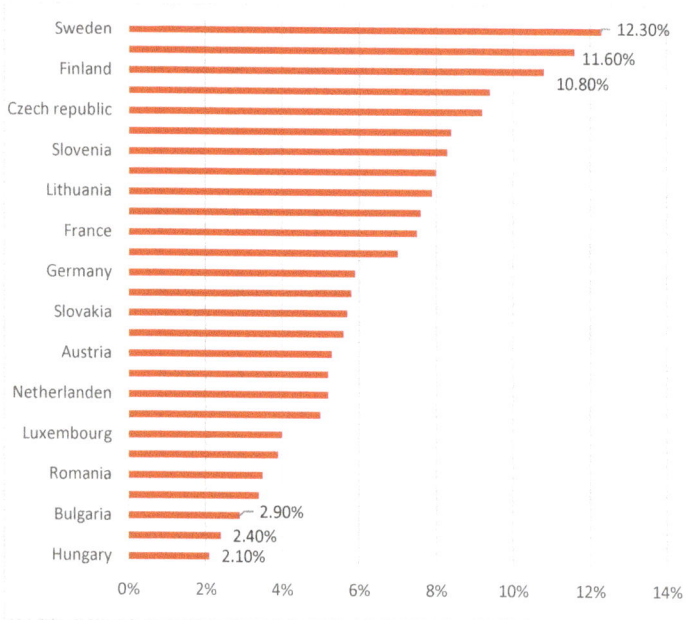

Figure 2. The labor force employed in the production of goods for environmental protection in EU countries in 2017. Source: based on processed data provided by Eurostat (2017).

Thus, the EU countries with the highest percentage of employees in the field of services and production of environmental goods (in terms of the total active population) were Sweden (12.3%), Denmark (11.6%), and Finland (10.8%), while the countries with the lowest percentage of employees in this area were Bulgaria (2.9%), Portugal (2.4%), and Hungary (2.1%).

Furthermore, Figure 3 reveals the status of the recycling rate of municipal waste in the EU countries in 2017.

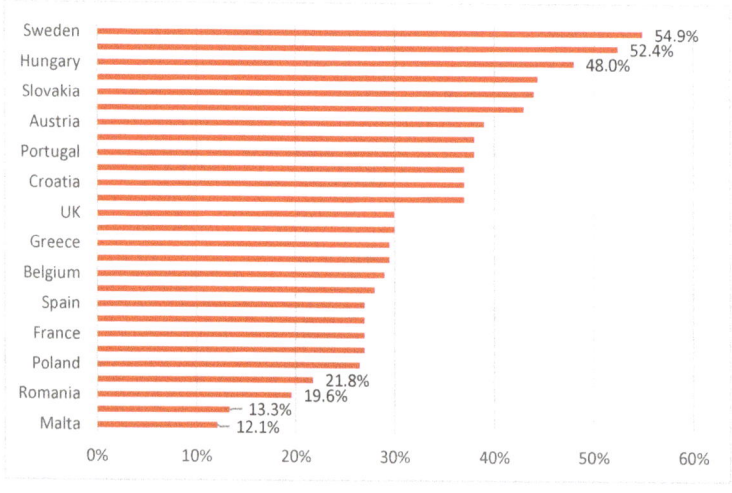

Figure 3. Recycling rate of the municipal waste in EU countries. Source: based on processed data provided by Eurostat (2017).

From the figure above, according to Eurostat, we could conclude that the EU countries with the highest recycling rate of municipal waste were Sweden with 54.9%, Denmark with 52.4%, and Hungary with 48%. The countries with the lowest recycling rates were Romania with 19.6%, Cyprus with 13.3%, and Malta with 12.1%.

Figure 4 illustrates the shares of innovative enterprises that brought new value-added and environmental benefits to the EU member states in the year 2017.

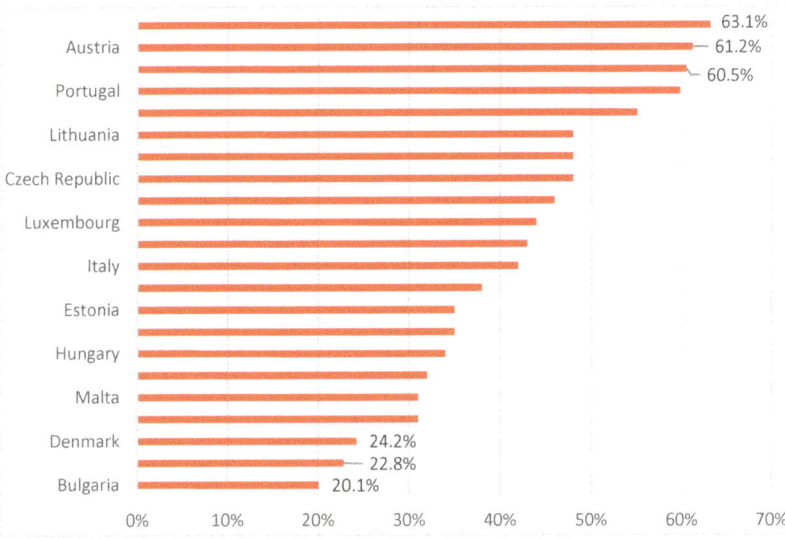

Figure 4. The share of enterprises that have brought innovative new products with environmental benefits to market in EU countries in 2017. Source: based on processed data provided by Eurostat (2017).

In the figure above, it can be seen that, in 2017, Germany ranked first (63.1%), followed by Austria (61.2%) and Finland (60.5%). In the same time, Denmark was positioned at the third-last level in the EU, with a share of 24.2% of companies which brought innovative new products with environmental benefits to the market, followed by Romania (22.8%) and Bulgaria (20.1%)

Figure 5 shows us the ranking among EU member states with respect to the use of green energy, calculated as a percentage of total energy consumption in 2017.

From this graph we could conclude that Nordic EU states were the countries with the highest green energy use, namely, Sweden (50.9%), Finland (42.1%), and Latvia (39.1%). On the opposite side were the following countries: the Netherlands (6.2%), Luxembourg (5.9%), and Malta (5.4%).

In conclusion, the descriptive analysis of the circular economy indicates that the Nordic countries were at the top of the rankings, with the highest degree of circular economy implementation, while the EU countries from Southeastern Europe were at the lowest places. One of the possible explanations comes from the environmental policies which were developed in the Nordic countries. For instance, the Nordic countries have implemented the "Circular Public Procurement in the Nordic Countries" (CIPRON), which is a process expected to provide conditions and criteria that would stimulate energy and material savings and closed material loops, in addition to spreading innovative solutions and creating markets for clean solutions. Another explanation could come from the fact that these countries have implemented alternative energy use in their countries. Moreover, according to the Eco-Innovation and Competitiveness Annual Report (2017), at the European level, Sweden, Finland, and Denmark ranked highest in the composite score (16 indicators arranged in 5 components) in particular due to their top performance in the components of eco-innovation inputs, activities, and outputs.

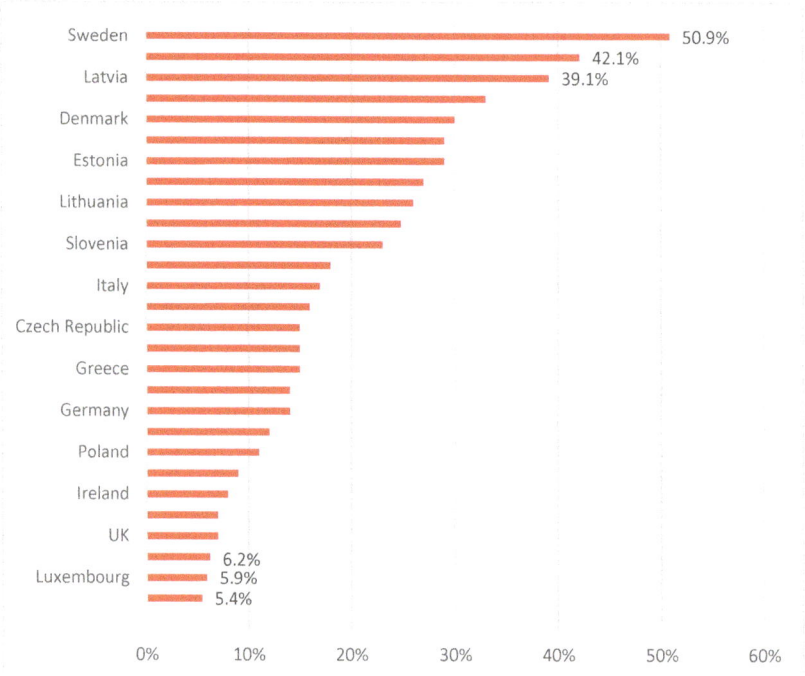

Figure 5. The green energy use in total energy consumption in EU countries, in 2017. Source: based on processed data provided by Eurostat (2017).

3. Results

Descriptive statistics related to the variables used in the analysis (i.e., mean, median, maximum, minimum, and standard error) are given in Table 2. The mean and median values of the descriptive statistics for the variables in the model in Table 2 are indicative of how close the data are to normal dispersion. In cases where the data have a standard normal distribution, the mean and median values approximate each other (Chang 2008). Table 2 also shows that the mean and median values of all variables were very close to each other. For this reason, it was assumed that all variables in the model were close to a standard normal distribution.

Table 2. Statistical description of variables in the model.

Variable	Mean	Median	Standard Deviation	N
Y	0.2014	0.1942	0.1043	27
X_1	8.9432	8.9753	3.3022	27
X_2	0.1520	0.2013	0.0834	27
X_3	0.6012	0.7021	0.1027	27
X_4	0.9021	0.8042	0.0702	27
X_5	0.8124	0.9324	0.0987	27

Source: Data analysis was performed by the author in EViews 10.0 (Eviews 2017).

A correlation table was used to investigate the existence of multicollinearity among the exogenous variables used in the model. Since the correlation coefficients in Table 3 are smaller than ±0.70, it was assumed that there were multicollinearity problems among the variables (Weinberg and Carmeli 2008).

Table 3. Correlation matrix.

Variable	Y	X_1	X_2	X_3	X_4	X_5
Y	1	0.702	0.539	0.616	0.632	0.792
X_1	0.702	1	0.028	0.034	0.112	0.109
X_2	0.602	0.034	1	0.052	0.106	0.078
X_3	0.598	0.028	0.048	1	0.079	0.081
X_4	0.623	0.112	0.106	0.079	1	0.102
X_5	0.792	0.109	0.078	0.081	0.102	1

Source: Data analysis was performed by the author in EViews 10.0.

F-test and Lagrange multiplier Breusch–Pagan test were performed to determine whether the method used in our analysis of the research model in Equation (1) was fixed effects, random effects, or pooled data, after reporting descriptive statistics for the variables used in the model.

The aim of using the F-test is to test the validity of the pooled model from the intended static panel data models against the fixed effect model (Urban 2015; Schmidheiny 2016). Restricted and unrestricted models are required to perform this test.

Restricted model:
$$Y_i = X\beta + u_i, i = \overline{1, N}. \tag{1}$$

Unrestricted model:
$$Y_i = X_i\beta_i + u;\ H_0 : \beta_i = \beta;\ H_0 : \beta_i \neq \beta.$$

If the null hypothesis (H_0) is not rejected; it will be $\beta_i = \beta$ in such a case, a classical model is accepted, and a solution is made by using the pooled data technique. Otherwise, the fixed effect model will be valid.

Table 4 shows the F-test statistical results. Hypothesis H_0 was accepted since the probability value was greater than the error according to these results. It was determined that the fixed effect model would not be suitable for our analysis.

Table 4. Fixed effect test.

F Statistics	3.56
F Stat. prob.	0.210

The Lagrange multiplier Breusch–Pagan test is used to make a choice between the pooled method and the random models (Block 2009). The hypothesis that the variance of random effects is zero is as follows:

$$H_0 : \sigma_u^2 = 0;\ H_1 : \sigma_u^2 \neq 0;$$

if the variance of the unit effects is zero, it indicates that the model will be analyzed with the pooled model. The results of the Breusch–Pagan test are shown in Table 5.

Table 5. Random effect test.

	Cross Section	Time	Both
Coefficient	32.23	54.324	89.65
Probability	0.089	0.567	0.035

Upon examining the test results in Table 5, hypothesis H_0 was accepted when the probability value was greater than 0.05. In this case, it was concluded that the random effect model would not be suitable for our analysis. Therefore, we could conclude that the pooling of the model in Equation (1) was appropriate.

The regression equation used to test the three statistical hypotheses was accomplished using the pooled least square method. This method was used to estimate performance-based economic growth and the use of circular economy at the EU level between 2008 and 2017.

For the regression analysis, the economic growth was considered as the dependent variable (Y) influenced by a set of four independent factors (regressors), which were the productivity of the resources (X_1), employment in production of environmental goods (X_2), recycling rate (X_3), and the market shares of the innovative enterprises which have brought innovative new products with environmental benefits to the market (X_4). Multiple linear regression analysis covered the following stages:

- Development of the regression model;
- Estimating model parameters; and
- Checking the accuracy of the results.

Analyzing the evolution of economic growth during 2008–2017 at the EU level, according to the independent variables, the following results were obtained for the multiple regression function using the multifactorial linear regression model (Table 6):

Table 6. Impact of resource productivity, recycling rate, consumption, environment innovation, and recycling rate of correlations on GDP per capita growth at the EU level.

Dependent Variable: GDP_CAPITA_GROWTH Method: Pooled least squares Sample: 2008–2017 Total panel observations: 270 GDP_CAPITA_GROWTH = C(1) + C(2) × ROD_OF_RES + C(3) × ENVIRON_EMPL + C(4) × REC_RATE + C(5) × ENVIRON_INNOV + C(6) × RENEWABLE				
	Coefficient	Std. Error	t-Statistic	Prob.
C	−1.9335	1.245	3.252243	0.0084
Prod_of_res	1.734502	1.430	3.58543	0.0045
Environ_Empl	1.168740	1.320	4.188850	0.0053
Rec_Rate	0.457692	1.012	4.821022	0.0047
Environ_Innov	0.503238	0.795	3.328985	0.0310
Renewable	0.687912	1.768	4.567883	0.0210
R-squared	0.798652	Mean dependent var		8.7892
Adjusted R-squared	0.824678	S.D. dependent var		0.6043
S.E. of regression	0.098723	Akaike info criterion		1.8023
Sum squared resid	1.408763	Schwarz criterion		1.7098
Log likelihood	119.3076	Hannan–Quinn criter.		1.6987
Durbin–Watson stat	2.088790			

According to the above table, the regression equation is:

$$Y = -1.9335 + 1.734X_1 + 1.168X_2 + 0.457X_3 + 0.503X_4 + 0.687X_5, \qquad (2)$$

where:

- Y = GDP per capita growth;
- X_1 = labor productivity;

- X_2 = labor force engaged production of environmental goods;
- X_3 = recycling rate of municipal waste;
- X_4 = the share of innovative enterprises that have brought innovative new products with environmental benefits to the market;
- X_5 = use of renewable energy.

Thus, according to Table 7, all three statistical assumptions were valid.

Table 7. Validation of the statistical hypotheses.

Hypothesis	Validated (Yes/No)
Hypothesis 1	Yes
Hypothesis 2	Yes
Hypothesis 3	Yes

A description of the regression variables can be observed in Table 2.

4. Discussion

In this chapter we discuss the factor analysis resulting in standardized pooled least squares (PLS). The method was used by the author to estimate the impact of circular economy on growth in the EU countries.

The relationship between CE level and economic growth has received attention in recent economic literature. Multiple linear regression model parameters used in this study were estimated by PLS, and the analysis was performed by EViews 10.0 software.

Analyzing the evolution of economic growth in the 27 EU member states in 2008–2017 through independent variables (i.e., GDP per capita, productivity of the resources, employment in the production of environmental goods, recycling rate of municipal waste, market share of the enterprises that have brought innovative new products with environmental benefits to the market, and renewable energy use), the following results were obtained through the analysis of multifactorial regression (Table 2): $Y = -1.9335 + 1.734X_1 + 1.168X_2 + 0.457X_3 + 0.503X_4 + 0.687X_5$, with standard error coefficients (1.245), (1.430), (1.012), (0.795), and (1.768).

Since R-squared was 0.7986 we concluded that 79.86% of the variability of the dependent variable was explained by the independent variables in the model. Additionally, the Durbin–Watson test indicated that there were no collinearity problems between the independent variables in the model since the value of statistical test was DW = 2.09, very close to 2, which leads to the conclusion that there was no autocorrelation of errors. The positive coefficient $\beta1$ confirms our expectations regarding the convergence between countries with low to high resource productivity. As expected, the recycling rate of municipal waste was significant and positive. Coefficients of predictor variables were also significant and positive, which means that the proxies used for the CE had a positive and significant impact on economic growth.

As can be seen from Table 6, the model estimation results were statistically significant at a significance level of 95% for all four independent variables in the model. The results of our analysis are consistent with the work of Puigcerver-Peñalver (2007), who developed a regression model for economic growth that is partly explained by environmental factors and CE.

The econometric model revealed that our analysis is valid and correctly specified and the environmental circular economy factors were significant indicators of economic growth in all 27 EU countries, given that the regression-estimated coefficients of the model were significantly different than zero and most of the variation of economic growth in EU countries was explained by the model. The results of our study are in line with recent papers of circular economy impact on economic growth (Preston 2012; Su et al. 2013; and Bocken et al. 2016).

The results of our analysis were consistent with the work of Sofian et al. (2017), which developed a regression model to explain the economic growth of EU member states, partly explained by environmental factors and innovation. The results also are connected to a study by Biber-Freudenberger et al. (2018), which highlighted that sustainability is not improved by a simple shift to renewable resources or materials. The authors argued that the productivity of the resources, environmental innovation, and recycling rates are important factors of economic growth and sustainable development.

5. Conclusions

The European Commission report on environmental policy indicates the increasing rates of resource reuse in EU countries. The level of implementation of the CE model requires constant and significant investments in the environmental infrastructure in order for EU countries to develop towards meeting the EU's environmental objectives.

Shortcomings were observed in the labor employed in the field of environmental protection and productivity of resources. Beyond the inventory of the current panorama of the implementation of CE at the EU level, this paper presents the advantages of using a conceptual model based on the efficient and responsible consumption of resources for sustainable economic growth.

The studies on EU developed economies have shown various benefits based on programs for the education of civil society in environmental protection, as well as making investments in infrastructure for collection, sorting, and recycling. The positive effects of implementing circular economy models increase the level of labor employed, municipal revenues, and the profit earned by entrepreneurs providing environmental infrastructure.

Nevertheless, the most important benefit of using the circular economy is felt individually. Making an analogy between extending products' life through reuse and our daily life, it could be observed how environmental and recycling factors are improving the quality of human life.

The multiple regression analysis carried out in this study reveals the impact of independent CE factors on dependent economic growth. At the same time, we conclude that the degree of innovation in the environment and the use of renewable energy play a greater role in terms of economic growth impact rate compared to the impact of GDP/capita and increasing human capital involved in renewable energy.

Our results confirm the statistical hypothesis, mainly related to the strong and significant effect of resource productivity on economic growth, confirming the European point of view that an increase of the resource productivity of 30% by 2030 may lead to a GDP growth of almost 1% (EU 2002).

This study could be of great importance for local, regional, and national public authorities of all EU countries involved in framing the legislative ground, as well as for enterprises, which could elaborate their business plans according to the predicted effects the implementation of circular economy may have on every member state.

The main limitation of this research is related to the time database used for the factor analysis, given that the calculation of macroeconomic indicators used in the multiple regression analysis covered a period of ten years. Thus, future research should be conducted for longer periods of time, which may provide a better panorama of the regression model applied for the macroeconomic indicators of the CE.

In conclusion, following our analysis, we can state that the economic model on economic growth was valid and accurately specified, and that the factors of renewable energy, productivity of the resources, recycling rate, environmental employment, and innovation were significant factors of economic growth at the EU level. This was because they registered significant values for the estimated coefficients, which were significantly different from zero, and because the model explained most of the variation in the economic growth of EU member states. This paper adds to the recent studies of the impact of CE on economic growth at the EU level (Geissdoerfer et al. 2017; Zink and Geyer 2017; Ghisellini et al. 2016).

Funding: This research received no external funding.

Conflicts of Interest: The author declares no conflict of interest.

References

Ayres, Robert. 1995. Life cycle analysis: A critique. *Resources, Conservation and Recycling* 14: 199–223.
Biber-Freudenberger, Lisa, Amit Kumar Basukala, Martin Bruckner, and Jan Börner. 2018. Sustainability Performance of National Bio-Economies. *Sustainability* 10: 2705. [CrossRef]
Block, Joern. 2009. *Long-Term Orientation of Family Firms: An Investigation of R&D Investments, Downsizing Practices, and Executive Pay*. Wiesbaden: Gabler, Available online: https://link.springer.com/book/10.1007/978-3-8349-8412-8 (accessed on 18 March 2019).
Blomsma, Fenna, and Geraldine Brennan. 2017. The emergence of circular economy: A new framing around prolonging resource productivity. *Journal of Industrial Ecology* 21: 603–14. [CrossRef]
Bocken, Nancy, Ingrid de Pauw, Conny Bakker, and Bram van der Grinten. 2016. Product design and business model strategies for a circular economy. *Journal of Industrial and Production Engineering* 33: 308–20. [CrossRef]
Brock, William, and Scott Taylor. 2005. Economic Growth and the Environment: A Review of Theory and Empirics. In *Handbook of Economic Growth*. vol. 1, pp. 1749–821. Available online: http://www.sciencedirect.com/science/article/pii/S1574068405010282 (accessed on 14 February 2019).
Browne, David, Bernadette O'Regan, and Richard Moles. 2009. Use of carbon footprinting to explore alternative household waste policy scenarios in an Irish city-region. *Resources, Conservation and Recycling* 54: 113–22. [CrossRef]
Cappa, Francesco., Jeffrey Laut, Oded Nov, Luca Giustiniano, and Maurizio Porfiri. 2016. Activating social strategies: Face-to-face interaction in technology-mediated citizen science. *Journal of Environment Management* 182: 374–84. [CrossRef]
Chang, Ni-Bin. 2008. Economic and policy instrument analyses in support of the scrap tire recycling program in Taiwan. *Journal of Environmental Management* 86: 435–50. [CrossRef] [PubMed]
Clodnitchi, Roxana, and Busu Cristian. 2017. Energy poverty in Romania–drivers, effects and possible measures to reduce its effects and number of people affected. Paper presented at International Conference on Business Excellence, Bucharest, Romania, March 21–22; vol. 11, pp. 138–45.
Cotae, Catrinel Elena. 2015. Regional performances in the context of a transition towards the circular economy: Structuring the assessment framework. *Ecoforum Journal* 4. Available online: http://ecoforumjournal.ro/index.php/eco/article/view/222 (accessed on 18 August 2017).
De Wolf, Catherine, Francesco Pomponi, and Alice Moncaster. 2017. Measuring embodied carbon dioxide equivalent of buildings: A review and critique of current industry practice. *Energy and Buildings* 140: 68–80. [CrossRef]
Eco-Innovation and Competitiveness Annual Report. 2017. Available online: http://eco.nomia.pt/contents/documentacao/kh0414991enn-002.pdf (accessed on 13 May 2019).
EU. 2002. Regulation (EC) No 2150/2002 of the European Parliament and the Council on Waste Statistics. Available online: http://www.kluwerlawonline.com/abstract.php?id=EELR2003019 (accessed on 23 December 2018).
Eurostat. 2017. Eurostat. Your Key to European Statistics. Available online: http://ec.europa.eu/eurostat (accessed on 24 September 2018).
Eviews. 2017. *User Guide*. version 10.0. Irvine: QMS Quantitative Micro Software, pp. 140–41.
Geissdoerfer, Martin, Paulo Savaget, Nancy Bocken, and Erik Jan Hultink. 2017. The Circular Economy—A new sustainability paradigm? *Journal of Cleaner Production* 143: 757–68. [CrossRef]
Geng Yong, Jia Fu, Joseph Sarkis, and Bing Xue. 2012. Towards a national circular economy indicator system in China: An evaluation and critical analysis. *Journal of Cleaner Production* 23: 216–24. [CrossRef]
George, Donald Ar, Brian Chi-ang Lin, and Yunmin Chen. 2015. A circular economy model of economic growth. *Environmental Modelling & Software* 73: 60–63.
Ghisellini, Patrizia, Catia Cialani, and Sergio Ulgiati. 2016. A review on circular economy: The expected transition to a balanced interplay of environmental and economic systems. *Journal of Cleaner Production* 114: 11–32. [CrossRef]

Gopal, Anandasivam, Manu Goyal, Serguei Netessine, and Matthew Reindorp. 2013. The impact of new product introduction on plant productivity in the North American automotive industry. *Management Science* 59: 2217–36. [CrossRef]

Grossman, Gene, and Alan Krueger. 1995. Economic growth and the environment. *The Quarterly Journal of Economics* 110: 353–77. [CrossRef]

Haas, Willi, Fridolin Krausmann, Dominik Wiedenhofer, and Markus Heinz. 2015. How circular is the global economy?: An assessment of material flows, waste production, and recycling in the European Union and the world in 2005. *Journal of Industrial Ecology* 19: 765–77. [CrossRef]

Kirchherr, Julian, Denise Reike, and Marko Hekkert. 2017. Conceptualizing the circular economy: An analysis of 114 definitions. Resources. *Conservation and Recycling* 127: 221–32. [CrossRef]

Lundvall, Bengt-Ake. 1996. *The Social Dimension of the Learning Economy*. Available online: https://pdfs.semanticscholar.org/202b/775ebcdbcaf8fd7c052f9a37a23776a3ea13.pdf (accessed on 7 May 2018).

Lyasnikov, Nikolay Vasilievich, Mikhail Nikolaevich Dudin, Vladimir Dmitriyevich Sekerin, Mikhail Yakovlevici Veselovsky, and Vera Grigoryevna Aleksakhina. 2014. The national innovation system: The conditions of its making and factors in its development. *Life Science Journal* 11: 535–38.

McDonough, William, and Michael Braungart. 2002. *Cradle to Cradle: Remaking the Way We Make Things*. New York: North Point Press.

Porter, Michael, and Claas Van der Linde. 1995. Toward a new conception of the environment-competitiveness relationship. *Journal of Economic Perspectives* 9: 97–118. [CrossRef]

Preston, Felix. 2012. *A Global Redesign? Shaping the Circular Economy*. London: Chatham House, Available online: https://www.bitcni.org.uk/wp-content/uploads/2014/11/bp0312_preston.pdf (accessed on 11 April 2019).

Puigcerver-Peñalver, Mari Carmen. 2007. The impact of structural funds policy on European regions' growth. A theoretical and empirical approach. *The European Journal of Comparative Economics* 4: 179.

Schmidheiny, Kurt. 2016. Panel Data: Fixed and Random Effects. Basel Universität. Available online: http://www.schmidheiny.name/teaching/panel2up.pdf (accessed on 18 December 2017).

Sjöström, Magnus, and Goran Östblom. 2009. FutureWaste Scenarios for Sweden Based on a CGEmodle. Working Paper 109, National Institute of Economic Research, Stockholm, Sweden. Available online: https://econpapers.repec.org/paper/hhsnierwp/0109.htm (accessed on 11 January 2019).

Sofian, Saudah, Mike Tayles, and Richard Pike. 2017. The implications of intellectual capital on performance measurement and corporate performance. *Jurnal Kemanusiaan* 4: 13–24.

Su, Biwei, Almas Heshmati, and Yong Geng. 2013. A review of the circular economy in China: Moving from rhetoric to implementation. *Journal of Cleaner Production* 42: 215–27. [CrossRef]

Urban, Markus. 2015. *The Influence of Blockholders on Agency Costs and Firm Value an Empirical Examination of Blockholder Characteristics and Interrelationships for German Listed Firms*. Berlin and Heidelberg: Springer.

Weinberg, Diana, and Abraham Carmeli. 2008. Exploring the antecedents of relationship commitment in an import–export dyad. In *New Perspectives in International Business Research*. Edited by Maryann Feldman and Grazia Santangelo. Bingley: Emerald, pp. 113–36.

Zink, Trevor, and Roland Geyer. 2017. Circular economy rebound. *Journal of Industrial Ecology* 21: 593–602. [CrossRef]

© 2019 by the author. Licensee MDPI, Basel, Switzerland. This article is an open access article distributed under the terms and conditions of the Creative Commons Attribution (CC BY) license (http://creativecommons.org/licenses/by/4.0/).

Article

The Development of the Health and Social Care Sector in the Regions of the Czech Republic in Comparison with other EU Countries

Erika Urbánková

Department of Economic Theories, Faculty of Economics and Management, Czech University of Life Sciences, Prague 16500, Czech Republic; urbankovae@pef.czu.cz

Received: 6 April 2019; Accepted: 29 May 2019; Published: 3 June 2019

Abstract: In this paper, the quantitative status of employees in the Health and Social Care sector in the Czech Republic is assessed, and the future development of the sector is predicted both for the Czech Republic as a whole, and for individual regions according to the NUTS3 classification. At present, labor market prognoses are created using the ROA-CERGE model, which includes the main professions in the Health and Social Care sector. This article expands the predictions by adding the regional level and using extrapolation of time series, and it identifies the regions important for the given sector and the labor force. The position of the Czech Republic with regard to selected professions in comparison with other countries of the European Union, i.e., its qualitative status, is also assessed in the paper. The following professions are assessed: general nurses and midwives (both with and without a specialization), physicians, and professional assistants. Healthcare workers do not manifest geographical mobility between regions and work primarily in the region where they live. Since the Czech Republic's accession to the EU, staff working in key professions have been able to work under comparable conditions in any of the member states. The workforce flow depends, among other things, on its qualitative representation in the given country. To find groups of European countries with similar characteristics of quantitative coverage in selected professions in the Health and Social Care sector, cluster analysis is used to identify homogeneous clusters of countries, as of 2016. Secondary data was obtained from the Czech Statistical Office (CZSO) and the Information System (ISA+) of the National Institute of Education (NIE).

Keywords: Health and Social Care; branches; NUTS3 regions; employment; professions; Czech Republic; European Union; prediction; cluster analysis

1. Introduction

In the labor market, the pairing process is key. There is an increasing discrepancy between the qualifications of the labor force and the qualification requirements of vacant jobs. In economic theory, this situation is called 'structural unemployment'. Structural unemployment arises when the supply of jobs for a certain type of profession or demanding a certain educational level does not correspond to the demand for this type of profession or the supply of workers with this educational level. The level of education of a potential or current worker or the field in which they graduate does not correspond to the qualification requirements of the position (Howell 2005; Karpíšek 1999; Samuelson and Nordhaus 2008). This discrepancy in the labor market arises primarily as a result of structural changes in the economy, where some sectors are expanding, and others are disappearing or declining. However, it is also created as early as during the actual process of education, where the fields of study generate an insufficient number of qualified people for the given profession or industry (Trhlíková et al. 2006; Brown et al. 2003; Redor 1999). Plesník et al. notes that continuously educating the workforce and increasing its spatial mobility are key to reducing structural unemployment (Plesník 2007; Holman 2001). However, this recommendation is suitable for fields with less demanding qualification requirements and no high specialization. Continuous education is based on the idea of so-called lifelong learning, i.e., the systematic development of individuals in all stages of their life. Experts distinguish three types of education, namely: formal, non-formal, and informal. Formal education is represented by the school education system, which is structured into primary, secondary (with or without graduation) and tertiary (higher education in a bachelor's, master's or doctoral degree) levels of education of an individual. Non-formal education is represented by the process of vocational education. Informal education is a form of self-education in which an individual acquires knowledge and skills through self-study (Bočková 2000; Severová 2011; Bills 2004; Bureš 2007). The economists from the Chicago School and their representative G.S. Becker (1964) do not portray education as a form of consumption, but rather consider the cost of education as a form of investment. The individual thus expects a certain return on their investment and anticipates future benefits (e.g., in the form of higher wages). Qualifications, or professional qualifications, are used to indicate a particular person's ability to practice a given profession. They indicate that a person possesses a certain level of professional training, education and knowledge in a particular field or profession (Schultz 1971). Qualifications consist of professional and general competencies. The qualification requirements of the Health and Social Care sector are high, and the individual professions are subject to requirements for increasing the level of formal education. The professional concentration of a sector shows to what degree a given sector is professionally homogeneous or heterogeneous, whereby the concentration is high if one or very few professions in the given sector have a dominant position in relation to the other professions. The average length of the period of education is an indicator that evaluates the individual levels of study by their difficulty expressed in terms of time; the time is the number of years spent studying at a given level of education (Mazouch and Fisher 2011). Economic theory views the elimination of disparities within regions differently, depending on the schools of economics. Liberals have favored an endogenous approach to regional development, whereby, in their opinion, it is not necessary to interfere in market processes. Under this conception, regional policy is governed by instruments that increase the mobility of labor forces and applies the approach of "the workers go where the work is" (Malinovský and Sucháček 2006; Maier and Tödtling 1998). As opposed to that, Keynesians have favored an exogenous approach to regional development and have emphasized the need for government interventions in the market mechanism, primarily in order to support market processes; which should be thereby regulated in such a way so that increases do not occur in disparities within the regions. The main instruments are thus steps to support the inflow of investments into problematic regions, i.e., the idea of "the work goes where the people are" (Myrdal 1957; Hirschman 1958). Marxists asserted the need for the planning and steering of spatial development, i.e., the utilization of a centrally planned economy, and ignoring market processes (Hampl 2001). The currently preferred modern transformed endogenous paradigm then places an emphasis primarily on the internal potential of individual regions, as well as the subjects found within them, and tends to pay rather less attention to divergent spatial development processes (Lundvall 1992; Florida 1995; Saxenian 1994; Piore and Sabel 1984). Regional disparities are to serve as a certain signal of the quality of the environment and the subjects within the individual areas of the country. According to the modern transformed endogenous approach to regional development, the key thing is primarily support for an environment in which small and midsize businesses, innovations, education and the development of human capital will thrive; as well as increasing the quality of local and regional institutions, maintaining competitiveness, and, at the same time, cooperation among entities (Blažek and Uhlíř 2002).

The Health and Social Care sector fits the theory of unbalanced growth. As a sector that is slower in productivity growth, it will tend to absorb more of the labor force. Given that public sector institutions generally provide services for which workforce productivity stagnates in time, there are reasonable concerns about their future development and existence. In this context, one of the most significant contributions of expert discussions is the concept of the model of unbalanced growth, which was put forward by the American professor of economics, William J. Baumol, in the

1960s. Baumol and Bowen's book, Baumol and Bowen 1966, is an analysis that follows the effect of delays in productivity. It is based on assuming the possibility of dividing the economy according to long-term changes in productivity. Activities manifesting long-term constant productivity tend to increase the income deficit, while the empirical data from various periods show that the income deficit grows faster than the general price level. The macroeconomic model of unbalanced growth assumes the division of every economy into two basic sectors, with the main criterion for the division being the technical infrastructure of production. Technologically-progressive sectors include economic activities that are characterized by long-term productivity growth. The reasons for this growth are mostly innovations (mainly technological), improvements in work skills, an increase in management efficiency, economies of scale, etc., while human labor is seen as a tool for achieving output. A sector that is not technologically progressive includes economic activities that are characterized by stagnating productivity, while human labor is usually the final output, or rather its integral part. In their article "Unbalanced Growth Revisited: Asymptotic Stagnancy and New Evidence", William J. Baumol together with Sue Anne Battey Blackman and Edward N. Wolf (Baumol et al. 1985) revised their original model. In terms of terminology, the activities with sporadic productivity growth or even with zero productivity growth are newly referred to as a stagnating sector. In addition to the stagnant and progressive sectors, there is the new concept of an asymptotically stagnant sector, which is characterized by its variability over time. As in the private service sector, the public service sector is characterized by stagnating labor productivity. This fact is supported by the claim that the same amount of workforce is always producing the same real value. A slight increase can then be explained, for example, by an absolute increase in the population of the given economy, as it can be assumed that production in the public service sector reflects the constant needs of consumers in the given economy. In the Health and Social Care sector, and in connection with the demand for workforce in this sector, demographic development and the aging of the population of the given country play a significant role, increasing the demand for services (labor and resources) in the area of healthcare. Another possible determinant is the growth of wages and salaries in the given country, and thus economic growth, when consumers increase their demand for services. Demographic development, the number of newborns, the number of students in the given field, the level of remuneration in the given profession, and the rigidity of the labor market primarily influence the labor supply in healthcare. Among other things, Baumol assumes that due to the effort to maintain the given production, there will be a significant spillover of labor capacity in favor of non-progressive activities while, at the same time, there will be considerable pressure in the form of a slowdown of total economic growth. Theoretically, the increase in labor capacity in the public service sector can be explained by population growth in the given country as well as by the increase in the gross domestic product, including real tax revenues, which can in return provide more services over time. As labor productivity in the public service sector has been constant over the long term, it is clear that production growth brings with it an increase in labor capacity.

In Figure 1a, the left axis represents the development of gross value added at constant prices (reference year 2010, in CZK millions) and the number of employed people, while the right axis shows the productivity curve, i.e., the ratio of the quantities from the left axis. The left axis in Figure 1b shows the development of production at constant prices (reference year 2010, in CZK millions) and the number of people employed, while the right axis shows the productivity curve, i.e., the ratio of quantities from the left axis (Český statistický úřad 2018c, 2018e). In the Czech Republic, the Health and Social Care sector shows decreasing productivity based on the calculation of gross value added per employee, and stagnating productivity based on the calculation of output per employee.

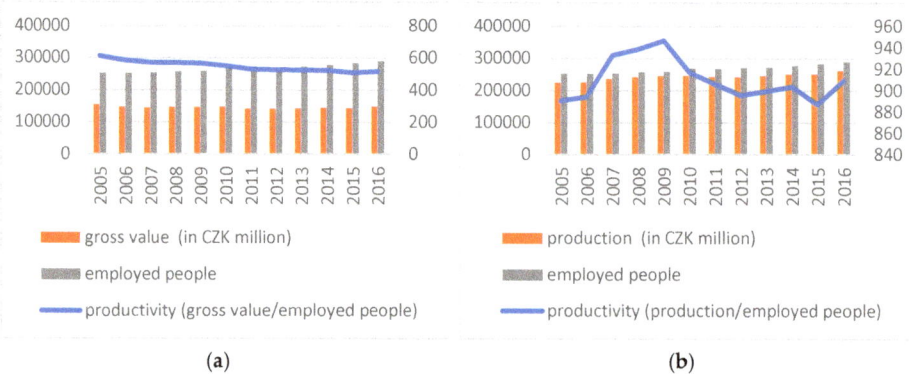

Figure 1. (a) Gross value added at constant prices (in CZK million, the number of employed people, productivity; (b) production at constant prices (in CZK million), the number of employed people, productivity; the Health and Social Care sector, the Czech Republic 2005–2016.

Demographic projections for the Czech Republic assume that in future years there will be a gradual increase in the number of people aged 65 and over. At the same time, the share of people in retirement in the total population is expected to exceed one fifth by 2020. Together with the ever-growing median age, this will place further demands for the increase of employment in health and social care. The total increase between 2008 and 2020 is anticipated to be very high—from 328,000 people in 2008 to 386,000 in 2020 (an increase of 18%). In the following years, the number of people employed in private health organizations will increase. Currently, the Czech Republic comes near to last in Europe in terms of this indicator. Among the most demanded professions will be primarily nurses, general practitioners and clinical doctors, dental experts, physiotherapists and psychology experts. In the short term, for example, a growing demand for staff in the areas of gynecology and obstetrics is anticipated. However, with the anticipated decline in the birth rate, this trend will soon be reversed in the coming years. The lack of other hospital support staff will remain a great problem in healthcare. Foreign workers will often be the only option for employers. In Figure 2 (Český statistický úřad 2018a), the secondary data from the Czech Statistical Office illustrate a negative demographic development and a natural decrease in the population of the Czech Republic between 1993 and 2005, while from 1983 on there were very low annual increases in the Czech Republic's population. The right axis depicts trends in the ageing index (the ratio of people over 65 to those aged 0–14, in %), which points to a significant rate of population aging. From 1989 to 2016, the aging index increased by 113%. This is caused by higher life expectancy and also by the declining birth rate.

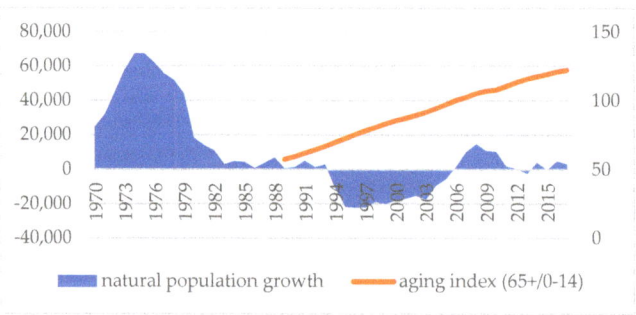

Figure 2. The trend in natural population growth between 1970 and 2016 (left axis) in the number of people; aging index trend between 1989 and 2016 (right axis) in %.

The systematic use of prognostic models has its place in the area of the future needs of the labor market. The European Union in particular, emphasizes the importance of forecasting the needs of the labor market as a tool for maintaining balance on the labor market and preventing structural unemployment. Predicting the necessary labor force is common practice and a necessary element of maintaining balance on the market in Germany, the Netherlands, Great Britain, the United States of America and France. In the Czech Republic, medium-term prognoses have a brief tradition and the model is currently at the development stage. The prognosis of labor market requirements for certain qualifications is conducted by the National Observatory of Employment and Training, with financial help from the Ministry of Labor and Social Affairs and the European Social Fund. In creating the model, the National Observatory of Employment and Training cooperates with the Research Institute for Labor and Social Affairs and the Centre for Economic Research and Graduate Education (CERGE-EI). For their prognosis, they used a model called ROA-CERGE. The model was taken from the Netherlands and modified for the environment of the Czech economy. Currently, selected economic sectors are analyzed in the Czech Republic (sector studies). Namely, studies have been conducted for the energy, electro-technical and ICT services sectors for the 2012–2020 period. The model further analyses the future development of economic sectors, anticipated future demand in the professional and educational structure and the anticipated future labor supply for individual professions and at certain levels of education. The supply side of the labor market is represented by a demographic model, or demographic projection, and it is mainly an estimate of the number of graduates in the professional, sector and qualification groups. The demand side consists of the number of vacancies for a certain level of education and particular professions. These vacancies are differentiated as newly created jobs that did not previously exist, and vacancies that are made available again due to existing staff entering their post-productive age. With regard to this differentiation, expanded demand (newly created jobs) and replacement demand (newly vacated jobs) are defined. Furthermore, the ROA-CERGE model works with "substitution demand", which represents demand for labor with a certain level of education for jobs that are characterized by this particular level of education. The quality-quantity model compares the anticipated supply with the anticipated demand on the labor market. The methods used in the model are: (a) monitoring of the labor market, (b) quantitative forecasts, (c) expert scenarios, (d) sectors studies, (e) shift-share analysis. Monitoring is carried out by means of statistical surveys conducted within the Labor Force Sample Survey. The quantitative forecasts are created using econometric and mathematical models; the statistical method of trend extrapolation is frequently used. Scenarios elaborated by experts are the pessimistic, baseline (realistic) and optimistic scenarios, which supplement the statistical and econometric models. Sector studies predict changes in sectors, expansion and downturn, employment trends and the number of jobs, the emergence and disappearance of individual professions in the sector; and they identify imbalance on the labor market in the individual sectors. Shift-share analysis identifies the main causes of the increase and decrease in employment in a single sector, a certain type of profession or a particular qualification level (Výzkumný ústav práce a sociálních věcí 2018). So far, this prognostic model does not provide outputs for forecasts on the regional level.

Practicing medical and non-medical health professions requires professional competence, which is defined by Act No. 96/2004 Coll. (amendment No. 201/2017), on conditions for the attainment and the recognition of professional qualifications to pursue non-medical healthcare related professions and to pursue other related activities and to provide medical care, and on the amendment of some other related acts. The four major professions in the Health and Social Care sector in the Czech Republic are classified according to CZ_ISCO: general nurses and midwives without a specialization (ISCO_322); general nurses and midwives with a specialization (ISCO_222); physicians and other healthcare specialists, with the exception of general nurses with a specialization (ISCO_22); and health professionals with the exception of general nurses without a specialization (ISCO_32) (Český statistický úřad 2018b; Národní ústav pro vzdělávání 2018a, Národní ústav pro vzdělávání 2018b, Národní ústav pro vzdělávání 2018c; Národní vzdělávací fond 2018):

- *General nurses and midwives without a specialization (ISCO_322)*

According to expert estimates, in the group General Nurses and Midwives without specialization; there will be changes in the structure of the sector by 2025, together with an internal change in professional structures within individual sectors. By 2025, the number of job positions is anticipated to increase by approximately 2000, i.e., by about 4% between 2016 and 2025; and at the same time, by 2025 approximately 8000 people are expected to leave their posts in this occupational group (primarily due to retirement). In total, it is anticipated that by 2025 approximately 10,000 vacancies will need to be filled. The largest number of people are employed in Prague. Of all people employed in this occupational group, 13% work in this region. About another 13% work in the Moravian-Silesian Region, over 12% in the Central Bohemian Region, almost 10% in the Ústí Region and over 8% in the Vysočina Region. Compared to other groups of professions, the geographical concentration of this occupational group is below average in the regions of the Czech Republic. The geographical exclusivity of this occupational group is low in the Czech Republic. The largest share this occupational group has in the total number of people employed regionally is in the Vysočina Region, where general nurses and midwives without specialization make up 1% of all people employed in the region. The majority of people in this occupational group work in the region they live in; regional mobility is very low in this occupational group. There are 4.1% of people who work in a different region, which is 1.9% less than the average for the whole of the Czech Republic. Make up 0.7% of the total number of employed people in the Czech Republic, and in the European Union this occupational group makes up 1.2% of all employed people. The country with the highest percentage of employed people in this group is Germany, which has almost 3.5% (followed by Finland, Slovakia, Belgium, Croatia, Austria, Italy, Luxembourg, Estonia, Slovenia and France, which are above the EU 28 average). The average length of the educational period for people employed in this profession is 12.3 years. People with secondary education make up the largest proportion in this occupational group (86%), and people with tertiary education make up 13% (of which 55% have a bachelor's degree). The qualification requirements of the jobs are 5.45 points in the Czech Republic and 5.46 points in the EU. People with a secondary education made up the largest proportion of this occupational group, and people with tertiary education made up the second largest part of this group.

- *General nurses and midwives with a specialization (ISCO_222)*

By 2025, an increase of approximately 6000 (i.e., about 10%) in the number of staff is anticipated for the occupational group General Nurses and Midwives with specialization, and by this time, approximately 19,000 people should have left their jobs (mainly to retire). In total, it is anticipated that by 2025 approximately 25,000 job vacancies will need to be filled. The largest number of people are employed in Prague. Of all people employed in this occupational group; 17% work in this region. Over 14% work in the South Moravian Region, almost 10% in the Moravian-Silesian Region, almost 9% in the Central Bohemian Region and over 7% in the South Bohemian Region. Compared to other occupational groups, the geographical concentration of this group is average in the regions of the Czech Republic. The geographical exclusivity of this occupational group is below average in the Czech Republic. The largest share this occupational group has in the total number of people employed regionally is in the South Moravian Region, where it makes up over 1% of all people employed in the region. 4.8% of people work in a different region, which is 1.1% less than the average for the whole of the Czech Republic. The share of general nurses and midwives with a specialization in the total number of employed people in the Czech Republic is 1.3%, and in the European Union it is 1.0% of all employed people. The country with the highest percentage of employed people in this category is Ireland, which has almost 3.4% (followed by Denmark, Sweden, Great Britain, Latvia, Malta, Poland, the Netherlands, Spain, Bulgaria, Belgium, the Czech Republic and Portugal, which are all above the EU 28 average). The average length of the educational period for all people employed in this occupational group is 13 years; the most frequent level of education is secondary education with graduation (68%), and 32% of people have tertiary education (of which 47% have a bachelor's degree). The qualification requirements of these jobs are 6.38 points in the Czech Republic and 6.37 points in the EU.

- *Physicians and other healthcare specialists, with the exception of general nurses with a specialization (ISCO_22)*

An increase in the number of jobs is anticipated in the occupational group Physicians and other healthcare specialists, with the exception of general nurses with a specialization by 2025. Between 2016 and 2025, their number should increase by approximately 7000, i.e., by approximately 10%; and at the same time, approximately 20,000 people are expected to leave their jobs. It is anticipated that by 2025, a total of about 27,000 job vacancies will need to be filled. The greatest number of people from this occupational group is employed in Prague of all people employed in this occupational group; 21% work in this region. Over 12% work in the South Moravian Region, almost 11% in the Central Bohemian Region, over 8% in the Ústí Region and almost 8% in the Moravian-Silesian Region. Compared to other occupational groups, the geographical concentration of this group is above average in the regions of the Czech Republic. The geographical exclusivity of this occupational group is below average in the Czech Republic. In the Pardubice Region, this occupational group has the largest share in the total number of people employed regionally, making up over 2% of all people employed in the region. There are 8.2% of people who work in a different region, which is 2.3% more than the average for the whole of the Czech Republic. Physicians and other healthcare specialists, with the exception of general nurses with a specialization make up 1.4% of the total number of employed people in the Czech Republic, and 1.8% of all employed people in the EU 28. The country with the highest percentage of employed people is Belgium, which has almost 3.7% (followed by Spain, Greece, Denmark, the Netherlands, Luxembourg, Germany, Sweden, Malta, France, Slovenia and the United Kingdom, which are above the EU 28 average). The average length of the educational period for all people employed in this occupational group is 17.1 years, and all people in the group have tertiary education. The qualification requirements of these jobs are 6.94 points in the Czech Republic and 6.96 points in the EU.

- *Health professionals with the exception of general nurses without a specialization (ISCO_32)*

An increase in the number of jobs is anticipated for the occupational group Health professionals with the exception of general nurses without a specialization by 2025. Between 2016 and 2025, their number should increase by approximately 2000, i.e., by approximately 3%; and at the same time, approximately 13,000 should leave their jobs (mainly to retire). It is anticipated that by 2025, a total of 15,000 job vacancies will need to be filled. Again, the largest number of people are employed in Prague. Of all people employed in this occupational group 21% work in this region. Over 12% work in the South Moravian Region, over 10% in the Central Boh emian Region, almost 10% in the Moravian-Silesian Region and almost 7% in the Ústí Region. Compared to other occupational groups, the geographical concentration of this group is above average in the regions of the Czech Republic. The geographical exclusivity of this occupational group is low in the Czech Republic. This occupational group has the largest share in the total number of people employed regionally in Prague, where it makes up over 1% of all people employed in the region. There is a total of 4.9% of people who work in a different region, which is 1.0% less than the average for the whole of the Czech Republic. Health professionals, with the exception of general nurses without a specialization make up 1.1% of the total number of employed people in the Czech Republic, and 1.5% of all employed people on average in the European Union. The country with the highest percentage of employed people in this category is Germany, which has almost 3% (followed by the Netherlands, Austria, Finland, Italy and Hungary, which are above the EU 28 average). The average length of the educational period for all people employed in this occupational group is 13.2 years; people with secondary education make up the largest proportion of this occupation group (62%), and people with tertiary education make up 36%. The qualification requirements of these jobs are 5.59 points in the Czech Republic and 5.56 points in the EU.

In their 1976 research report to the European Commission, "The Potential for Substituting Manpower for Energy", Walter Stahel and Genevieve Reday sketched the vision of an economy in loops (or circular economy) and its impact on job creation, economic competitiveness, resource savings, and waste prevention (Cradle to Cradle 2018). The report was published in 1982 as the book Jobs for Tomorrow: The Potential for Substituting Manpower for Energy (Korhonen et al. 2018). Promoting a circular economy was identified as national policy in China's 11th five-year plan starting in 2006 (Zhijun and Nailing 2007). In January 2012, a report was released entitled Towards the Circular Economy: Economic and business rationale for an accelerated transition. The report, commissioned by the Ellen MacArthur Foundation and developed by McKinsey & Company to consider the economic and business opportunity for the transition to a restorative, circular model. The report details the potential for significant benefits across the EU (Ellen MacArthur Foundation 2012). According to the European Commission, moving from a linear to a circular economy means strengthening Europe's competitiveness, reducing dependence on primary raw materials and creating jobs. The fundamentals of this concept can be found in Cradle to Cradle: Remaking the Way We Make Things from the German chemist Michael Braungart and American architect William McDonough. The great challenge of circular economy is to overcome the linear "taking, doing and liquidating" economic model (Govindan and Hasanagic 2018). According to Kirchherr et al. (2017), a circular economy is most frequently depicted as a combination of reduce, reuse, and recycle activities. This regenerative approach is in contrast to the traditional linear economy, which has a 'take, make, dispose' model of production.

Section 2 describes Materials and Methods. In Section 3; Results, the development of employed people is further predicted for the Czech Republic and its regions and the usability of the workforce in key professions in the sector is compared across the European Union. Section 4 contains the Discussion.

2. Materials and Methods

For the prediction of values and the extension of time series, the method of double exponential time series smoothing is used, due to their nature. For the double exponential smoothing type, a linear trend is assumed in the short sections. The exponential smoothing method uses all past observations of the time series for smoothing and thus has an adaptive approach to the trend component, and the underlying idea is that the importance of the observations decreases towards the past. To estimate the parameters, the modified least squares method is used. Double exponential smoothing is an adaptive method for estimating the parameters of a polynomial trend line for non-seasonal time series using only a single equalization constant a. Double exponential smoothing is used when it is possible to assume that in a short period, the trend line will be linear. The method of exponential smoothing is based on the application of the method of the weighted least squares to all available observations in the given time series, with weight of the individual observations decreasing exponentially towards the past. However, the hypothetical extension of the time series to the past is justified, as it allows a significant simplification of the formula for the calculation of the smoothed values and forecasts. In all variants of exponential smoothing, it is assumed that the smoothed time series has a shape $T_t = \beta_0 + \beta_1 t$. When applying the method of double exponential smoothing, it is assumed that the trend of the examined time series is linear over short time periods. By estimating parameters β_0 and β_1, using the method of least squares (Cipra 1986).

$$\text{Min} \sum_{t=1}^{n} [y_t - \hat{y}_t]^2 = \text{Min} \sum_{t=1}^{n} [y_t - \beta_0 - \beta_1 . t]^2 \tag{1}$$

The method of exponential smoothing is based on all previous observations, while their weight decreases towards the past according to the exponential function. Alfa α is the smoothing constant in the (0;1) interval (Cipra 1986).

$$\sum_{i=0}^{n} [y_{t-1} - \beta_0 - \beta_1(-i)]^2 \alpha^i \tag{2}$$

The intensity of forgetting/filtering expressed by alfa is determined based on the character of the time series. A value of α is sought for which MAPE is smallest, possibly MSE. The confidence interval is the area around each predicted value, into which 95% of the future value falls, with a lower confidence interval representing a higher accuracy of the estimate. Residual statistics are available for assessing the accuracy of the selected trend function, and the accuracy of the selected trend function is assessed according to the mean absolute percentage error (MAPE):

$$\sum \left(\frac{|y_t - T_t|}{y_t} \right) \times \frac{100}{n} \qquad (3)$$

which is suitable for comparing different time series due to its non-dimensionality (Artl et al. 2002).

Cluster analysis is used to find groups of similar countries; its aim is to examine whether it is possible to identify the differences between the monitored cases on the basis of selected economic factors. Cluster analysis is a multidimensional statistical method that can be used to identify a group of similarly behaving objects and group them into clusters. Objects inside of clusters are the most homogeneous, and objects belonging to different clusters are the most heterogeneous. Common types of cluster techniques include hierarchical clustering and the K-Means method. Hierarchical clustering begins with n clusters, where each observation forms a separate cluster and ends with one cluster that includes all observations. In each step, the two closest observations or observation clusters are merged into one new cluster. The clustering procedure is captured by a special tree graph, or so-called dendrogram, which shows the individual steps of hierarchical clustering, including the distances at which the individual clusters (or observations) were merged. The dendrogram is also used to present the results. The starting point for clustering is to determine the method of expressing the similarity (distance) of individual cases. The squared Euclidean distance was chosen as the distance metric in the presented work. This is the standard distance metric used in geometry, which is generalized to multidimensional data. Ward's method, which is based on the creation of clusters with the highest possible internal homogeneity, and is a hierarchical clustering method recommended by many experts, was used for the clustering itself. One of its advantages is that the chaining of clusters does not occur, as it does, for example, in the case of the nearest neighbor and farthest neighbor methods. Ward's method is based on the analysis of variance, and it selects and merges clusters with a minimum sum of squares. For the purposes of this work, hierarchical clustering is only used to determine the number of clusters that have a tendency to form naturally.

Ward's method:

$$\Delta C = \sum_{i=1}^{G} \sum_{j=1}^{n} (x_{gij} - v_{gj})^2 - \sum_{i=1}^{A} \sum_{j=1}^{n} (x_{aij} - v_{aj})^2 - \sum_{i=1}^{B} \sum_{j=1}^{n} (x_{bij} - v_{bj})^2 \qquad (4)$$

where: x_{gij} is the value of the *ith* element of cluster G, G is the number of elements of this cluster, v_{gj} is the average value of the *jth* clustered variable G, etc. (Meloun et al. 2011; Hebák et al. 2007).

The process of the cluster analysis is as follows (Hebák et al. 2007):

(a) Selection of input database, data type, objects and the variables of the analyzed objects.
(b) Selection of the types of variables, type of useful variables in data.
(c) Determination of the optimum number of clusters.
(d) Selection of the clustering method (simple average, group average, centroid-based clustering, clustering using the nearest-neighbor or farthest-neighbor chain algorithms, median method, Ward's method). For the purposes of this work, Ward's method was selected.

(e) Selection of distance used: Euclidian metric (geometrical distance), Hamming/Manhattan distance, Mahalanobis distance or the generalized Minkowski distance. For the purposes of this paper, the squared Euclidian distance has been used:

$$D_E(x, y) = \sum_{i=1}^{n} (x_i - y_i)^2 \qquad (5)$$

where: x is the current value of the component (variable) in the matrix, and y represents the number of factors (Meloun et al. 2011; Hebák et al. 2007).

(a) The process of linking and clustering, during which the calculation of distances between objects and clusters and the creation of a dendrogram can be carried out: by the method of hierarchical clustering, k-means clustering, k-medoids clustering, and fuzzy clustering.
(b) Interpretation of the dendogram and the distance matrix.

After the content of the clusters is determined, the K-Means technique is used following hierarchical clustering (which determines the quantitative number of clusters). This technique is suitable for clustering a larger number of objects according to selected variables. It works on the principle of minimizing the sum of the squares of deviations from the mean, which precisely coincides with the Ward criterion G1. The K-Means method is an algorithm based on the shifting of objects between clusters. It is a nearest centroid method that provides only one solution for a number of required clusters, and the objects are clustered based on the smallest distance between the object and the centroid of the cluster (Hebák et al. 2007). Meloun et al. (2011) note that it is necessary to select initial partitioning into k clusters, whereby the number of clusters required is obtained based on the results of hierarchical clustering. The classification according to which individual professions are distinguished in the analyses is called CZ-ISCO, which determines the occupational structure of the workforce and is based on the methodological principles of international classification ISCO-08, which was created by the International Labor Organization (ILO) (Český statistický úřad 2018a).

3. Results

3.1. Data

The trend for the number of people employed in the Health and Social Care sector in the Czech Republic from 1993 to 2016 (in thousands of people) is shown in Figure 3 (Český statistický úřad 2018e). The time series has a increasing linear trend, and the double exponential smoothing method is thus used. Considering its length of 23 observations, this time series is extended by five periods until 2021. The prediction of the trend for the given variable is shown in orange and contains confidence intervals (95%). The predicted series for the whole of the Czech Republic shows a slightly increasing trend in the short-term horizon. The MAPE value is 0.02, so the trend was appropriately selected. The prediction results for the individual years are as follows: 2017—363.59 thousand people; 2018—367.73 thousand people; 2019—371.87 thousand people; 2020—376.01 thousand people; 2021—379.33 thousand people. The sector thus has an expansionary trend.

The Czech Republic is further disaggregated into individual NUTS3 regions. The number of people employed in the individual regions in the Health and Social Care sector is monitored. The regions with the following designations enter the analysis: the Capital City of Prague (PHA), Central Bohemian Region (STRC), South Bohemian Region (JIHC), Olomouc Region (OLM), Zlín Region (ZLN), Plzeň Region, Hradec Králové Region, Liberec Region, Karlovy Vary Region (KAR), Ústí nad Labem Region (UST), Pardubice Region (PAR), Vysočina Region (VYS), South Moravian Region (JIHM), and the Moravian-Silesian Region (MOSL).

The number of people employed in the monitored sector varies in the different regions of the Czech Republic, mainly because of the size of the individual regions and the representation of health institutions. As can be clearly seen in Figure 4 (Český statistický úřad 2018e), the Capital City of Prague has the largest absolute number of people employed in the sector in the entire monitored period 1993–2016 (46,200 people in 2016) and is followed by the Moravian-Silesian Region (44,800 thousand people in 2016), the South Moravian Region (40,300 people in 2016), and finally, the Central Bohemia Region (45,100 people in 2016). Conversely, the Liberec Region (15,600 people in 2016) and the Karlovy Vary Region (13,200 people in 2016) had the lowest numbers of people employed in the sector in the monitored period. The other regions each have between 15,000–25,000 people in the entire monitored period. On the basis of this data, four major regions of employment in the given sector in the Czech Republic can be identified in the long-term horizon—the Capital City of Prague, the Moravian-Silesian Region, the South Moravian Region, and the Central Bohemian Region.

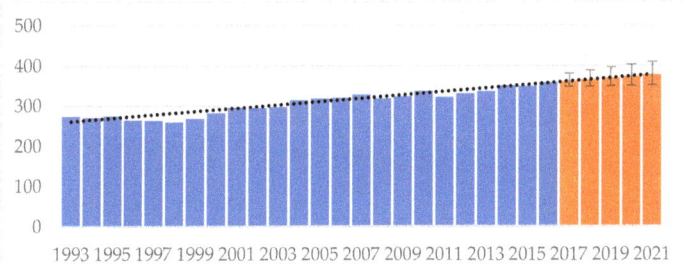

Figure 3. The trend for the number of people employed in the Health and Social Care sector in the Czech Republic (in thousands of people) 1993–2016, with a prediction 2017–2021.

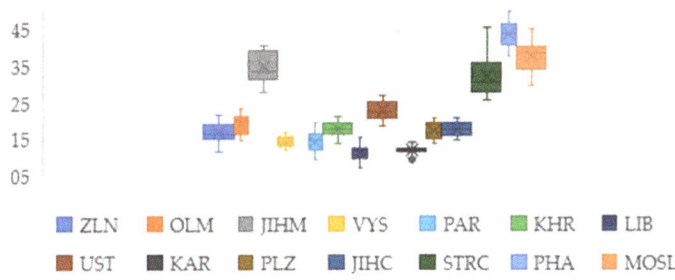

Figure 4. Boxplot of the number of people employed in the Health and Social Care sector in the regions of the Czech Republic in thousands of people (median for 1993–2016).

However, if we convert the absolute values in individual years to relative values, in the form of year-on-year changes, and use chain indices, i.e., the year-on-year rates of change, the regional trends for the given variable in the Health and Social Care sector show a different course. The boxplots in Figure 5 show the median of chain indices (year-on-year changes). The time series was divided into two monitored periods (namely 1993–2016 and 2004–2016) so that the development trend over the last decade is taken into account for the monitored variable. For the entire monitored period of 1993 to 2016 (Figure 4), the region with the highest rate of growth in the number of employed people is the Liberec Region, which has long had the highest rate of growth (average +3.81%, median +2.86%), followed by the Pardubice Region (average +2.84%, median +2.89%), the Moravia-Silesia Region (average +2.05%, median +2.46%) and, finally, the Central Bohemian Region (+1.95% average, median +1.92%). By shortening the time series to the period 2004–2016, it can be seen which region has a high rate of growth compared to the other regions in recent years, when the economic recession was manifested. If the time series is shortened, it can be said that the regions with the highest rate of growth are the Central Bohemian Region (average +1.96%, median +4.45%) and the Liberec Region (average + 2.15%, median +4.36%). Both regions had a significantly higher rate of growth in the monitored period than the rest of the regions, and their year-on-year declines were not as steep as they were for the other regions. Year-on-year declines of both expanding regions were recorded only in 2008 and 2012, and they had a significant rate of growth in the following years (the most significant year-on-year growth occurred in 2009, 2013, and 2016 for the Central Bohemian Region and in 2008, 2013, 2015 and 2016 for the Liberec Region). It can thus be said that these two regions are the most expanding regions in the Health and Social Care sector, regardless of the economic recession. Other important regions in the last decade are the Vysočina Region, the Ústí nad Labem Region and the Moravian-Silesian Region.

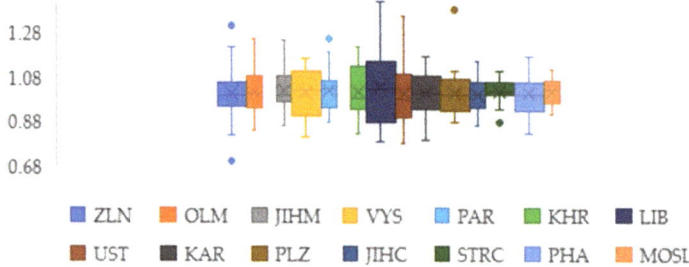

Figure 5. Boxplots of people employed in the Health and Social Care sector in the regions of the Czech Republic and median chain indices for the period 1993–2016.

The four important regions with the highest number of employees (PHA, MOSL, STRC and JIHM), the two regions with the lowest number of employees (LIB, KAR), and the two regions with the highest employment growth rate were thus selected for the prediction. The time series extrapolation is applied to Prague, the Moravian-Silesian Region, the South Moravian Region (the highest number of employed people), the Central Bohemian Region (the fourth highest number of employed people and the second highest growth rate), the Liberec Region (with the second lowest number of employed people and the highest growth rate) and the Karlovy Vary Region (with the lowest number of employed people.). The time series for the period 1993–2016 is extended by five periods, i.e., from 2017 to 2021.

3.2. Prediction

Double exponential smoothing was used to predict the values. From the results (Table 1), it is clear that all the regions will have a growing trend until the year 2021.

Table 1. The predicted values for the number of people employed in the Health and Social Care sector in selected regions of the Czech Republic (in thousands of people).

	PHA	STR	MOSL	JIHM	LIB	KAR
2017	473,828	4,660,286	431,965	4,094,536	1,381,689	1,340,019
2018	4,764,795	4,881,589	4,369,987	414,622	1,403,482	1,349,023
2019	4,791,309	5,102,892	4,420,324	4,197,903	1,425,275	1,358,028
2020	4,817,824	5,324,195	4,470,661	4,249,587	1,447,067	1,367,032
2021	4,844,339	5,545,498	4,520,998	430,127	1,468,816	1,376,037
basic indices (2021/2016)	1.022383	1.189948	1.046612	1.05049	1.063058	1.026879
percent (basic indices)	2.238344	18.9948	4.661208	5.049026	6.305833	2.687872
MAPE	0.069574	0.052092	0.061275	0.042384	0.106833	0.075384

In the past, the Capital City of Prague has had large fluctuations in the number of employed people, and its predicted growth is slight (2.24% increase over the entire predicted period), while the prediction for the Central Bohemian Region shows a rapid rate of growth (19%); this region could assume a dominant position as early as the year 2018. The Moravian-Silesian Region has the third highest number of employees and the fourth highest rate of growth (a 4.7% increase over the entire predicted period). The South Moravian Region has the fourth highest number of employed people and the third highest rate of growth (5% increase over the entire predicted period). The Liberec Region has the second lowest number of employees and the second highest rate of growth (6.3% increase over the entire predicted period). The Karlovy Vary Region has the lowest number of employees and its predicted values rise only slightly (2.7% increase over the entire predicted period).

3.3. Cluster Analysis of Homogeneity

A cluster analysis, which groups the countries into homogeneous clusters and identifies the differences between the clusters, is used to compare the Czech Republic with other EU countries and for the quantitative representation of the selected professions (the qualitative aspect) in the Health and Social Care sector.

The declared variables entered into the analysis are the following (Český statistický úřad 2018d):

- Employees in sector Q as a share of the total number of employed people in the country (EMPL_Q) in %,
- Share of employed physicians and other healthcare specialists, with the exception of general nurses with a specialization (ISCO_22) in %,
- Share of employed general nurses and midwives with a specialization (ISCO_222) in %,
- Share of employed health professionals, with the exception of nurses without a specialization (ISCO_32) in %,
- Share of employed general nurses and midwives without a specialization (ISCO_322) in %.

Given that the indicators have the same units of measurement, it was not necessary to standardize the data before entering it into cluster analyses by using the Z-score transformation, and the data entering the cluster analysis are thus percentage values. The year 2016 was studied. In the analysis, the EU countries are designated as follows: Belgium BE, Bulgaria BG, Czech Republic CZ, Denmark DK, Germany DE, Estonia EE, Ireland IE, Greece EL, Spain EC, Croatia HR, Italy IT, Cyprus CY, Latvia LV, Lithuania LT, Hungary HU, Malta MT, Austria AT, Poland PL, Portugal PT, Romania RO, Slovenia SI, Slovakia SK, Finland FI, Sweden SE, United Kingdom UK, Luxembourg LU, France FR, and the Netherlands NL.

The squared Euclidean distances were determined for the individual countries using Ward's method. Based on the analysis of the metric distances and the dendrogram evaluation, the creation of two very important clusters can be recommended. Two clusters are thus left for the K-Means clustering technique, and the differences are identified. Data in Figure 6 are from Český statistický úřad 2018d, own processing in program STATISTICA.

Cluster 1 (Italy, France, Croatia, Hungary, Finland, Slovakia, Estonia, Lithuania, Greece, Luxembourg, Slovenia and Germany) has the highest share of people employed in the professions of general nurse and midwife without a specialization ISCO_322 (1.7%); conversely, it has a low share of people employed in the professions of general nurse and midwife with a specialization ISCO_222 (0.48%). The share of employed health professionals is very close to the second cluster and makes up 1.15%, and the share of employed physicians is almost homogeneous for all countries and makes up 1.7%.

Cluster 2 (Belgium, Netherlands, Bulgaria, Cyprus, Spain, Poland, Portugal, the Czech Republic, Malta, Romania, Denmark, Sweden, Latvia, Great Britain and Ireland) has the lowest share of people employed in the professions of general nurse and midwife without a specialization ISCO_322 (0.3%); conversely, it has a high share of people employed in the professions of general nurse and midwife with a specialization ISCO_222 (1.6%). The share of employed health professionals is very close to the first cluster and makes up 1%, and the share of employed physicians is almost homogeneous for all countries and makes up 1.7%.

For a more detailed international analysis, it would be possible to create up to eight individual clusters for the given countries. However, due to the number of variables, the two basic clusters with the longest metric distance were selected.

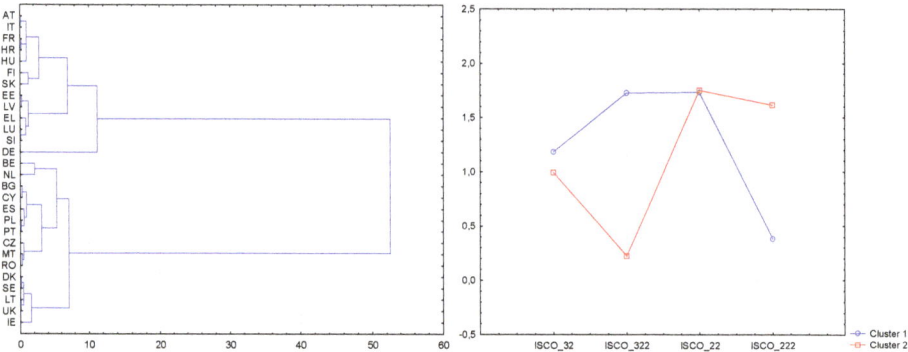

Figure 6. Dendrogram of hierarchical clustering (**left**) and the K-Means method (**right**).

4. Discussion

The aim of this paper is to evaluate the quantitative status of people employed in the Health and Social Care sector in the Czech Republic and individual NUTS3 regions, with a prediction of the sector's future development until 2021. At present, labor market prognoses are created using the ROA-CERGE model, which includes the main professions in the Health and Social Care sector. This article expands the predictions by adding the regional level and using extrapolation of time series, and it identifies the regions important for the given sector and the labor force. It has the partial objective of evaluating the position of the Czech Republic with regard to the selected professions compared with other EU countries, i.e., the qualitative status. In this paper, first the employment trends in the Health and Social Care sector in the Czech Republic and its individual regions were analyzed for the period 1993–2016.

The Health and Social Care sector fits the theory of unbalanced growth. In the Czech Republic, the Health and Social Care sector shows decreasing productivity based on the calculation of gross value added per employee, and stagnating productivity based on the calculation of output per employee. Demographic projections for the Czech Republic assume that in future years there will be a gradual increase in the number of people aged 65 and over. At the same time, the share of people in retirement in the total population is expected to exceed one fifth by 2020. Together with the ever-growing median age, this will place further demands for the increase of employment in health and social care. Regions that are expanding in terms of the rate of growth of the number of employed people, regions that employ the largest number of people in the given sector, and regions that employ the lowest number of people in the given sector were identified. The time series extrapolation was applied to the following regions:

- The Capital City of Prague has the highest number of employees in the entire period under study (46.2 thousand people in 2016);
- The Moravian-Silesian Region has the second highest number of employees in the entire under study period (44.8 thousand people in 2016) and the third highest growth rate over the entire period 1993–2016 (average +2.05%, median + 2.46%);
- The South Moravian Region has the third highest number of employees in the entire period under study (40.3 thousand people in 2016);

- The Central Bohemian Region has the fourth highest number of employees in the entire period under study (45.1 thousand people in 2016), the fourth highest growth rate in the entire period under study (+1.95%, +1.92%), and the second highest growth rate (average +1.96%, median +4.45%) in the shortened period 2004–2016;
- The Liberec Region has the second lowest number of employees in the entire period (15.6 thousand people in 2016) and the highest growth rate in both the entire period (average +3.81%, median +2.86%) and the shortened period 2004–2016 (mean +2.15%, median +4.36%);
- The Karlovy Vary Region has the lowest number of employees in the entire period under study (13.2 thousand people in 2016).

Double exponential smoothing was used to predict the values. From the results (specified in Table 1), it was clear that all the regions have a growing trend until the year 2021. In the past, the Capital City of Prague has had large fluctuations in the number of employed people, and its predicted growth is slight (2.24% increase over the entire predicted period), while the prediction for the Central Bohemian Region shows a rapid rate of growth (19%); this region could assume a dominant position as early as the year 2018. The Moravian-Silesian Region has the third highest number of employees and the fourth highest rate of growth (4.7% increase over the entire predicted period). The South Moravian Region has the fourth highest number of employed people and the third highest rate of growth (5% increase over the entire predicted period). The Liberec Region has the second lowest number of employees and the second highest rate of growth (6.3% increase over the entire predicted period). The Karlovy Vary Region has the lowest number of employees, and its predicted rise is slight (2.7% increase over the entire predicted period). In the Czech Republic, there is a growing demand for professionals in the Health and Social Care sector due to the negative demographic trends and the ageing of the population.

The four major professions in the Health and Social Care sector in the Czech Republic are classified according to CZ_ISCO (Český statistický úřad 2018b): general nurses and midwives without a specialization (ISCO_322); general nurses and midwives with a specialization (ISCO_222); physicians and other healthcare specialists, with the exception of general nurses with a specialization (ISCO_22); and health professionals with the exception of general nurses without a specialization (ISCO_32). According to expert estimates, in this groups of professions there will be changes in the structure of the sector by 2025, together with an internal change in professional structures within individual sectors. The demand for these professions will outstrip supply in the future. The majority of people in this occupational groups work in the region they live in; regional mobility is very low in this occupational groups. The differences and the homogeneity in the member states of the European Union in terms of the share of employed people in selected important professions in the Health and Social Care sector were also examined. In 2016 they formed clusters:

- The first cluster consisted of Austria, Italy, France, Croatia, Hungary, Finland, Slovakia, Estonia, Lithuania, Greece, Luxembourg, Slovenia and Germany: they had the highest share of people employed as General Nurses and Midwives without a Specialization. Make up 0.7% of the total number of employed people in the Czech Republic, and in the European Union this occupational group makes up 1.2% of all employed people. The country with the highest percentage of employed people in this group is Germany, which has almost 3.5% (followed by Finland, Slovakia, Croatia, Austria, Italy, Luxembourg, Estonia and Slovenia, which are above the EU 28 average. People with a secondary education made up the largest proportion of this occupational group, and people with tertiary education made up the second largest part of this group. The total level of qualification requirements in this profession was 5.45 points in the Czech Republic (on an eight-point scale) and 5.46 points in the European Union. The average length of the educational period for this occupation group was 12.3 years.
- The second cluster consisted of Belgium, the Netherlands, Bulgaria, Cyprus, Spain, Poland, Portugal, the Czech Republic, Malta, Romania, Denmark, Sweden, Latvia, Great Britain and Ireland, which have a high proportion of people employed as General Nurses and Midwives

with a Specialization. The share of general nurses and midwives with a specialization in the total number of employed people in the Czech Republic is 1.3%, and in the European Union it is 1.0% of all employed people. The country with the highest percentage of employed people in this category is Ireland, which has almost 3.4% (followed by Denmark, Sweden, Great Britain, Latvia, Malta, Poland, the Netherlands, Spain, Bulgaria, Belgium, the Czech Republic and Portugal, which are all above the EU 28 average). The total level of qualification requirements in this profession was 6.38 points (on an eight-point scale) in the Czech Republic and 6.37 points in the EU 28. The average length of the educational period of this occupational group was 13 years.

- Both clusters (all countries) show a small difference in the share of employed Health Professionals. Health professionals, with the exception of general nurses without a specialization make up 1.1% of the total number of employed people in the Czech Republic, and 1.5% of all employed people on average in the European Union. The country with the highest percentage of employed people in this category is Germany, which has almost 3% (followed by the Netherlands, Austria, Finland, Italy and Hungary, which are above the EU 28 average). The total level of qualification requirements of the jobs was 5.69 points in the Czech Republic (on an eight-point scale) and 5.66 points in the European Union. The average length of the educational period for this occupational group was 13.2 years.
- Both clusters (all countries) have a homogeneous share of employed Physicians and Other Specialists. Physicians and other healthcare specialists, with the exception of general nurses with a specialization make up 1.4% of the total number of employed people in the Czech Republic, and 1.8% of all employed people in the EU 28. The country with the highest percentage of employed people is Belgium, which has almost 3.7% (followed by Spain, Greece, Denmark, the Netherlands, Luxembourg, Germany, Sweden, Malta, France, Slovenia and the United Kingdom, which are above the EU 28 average). The total level of qualification requirements in this profession was 6.94 points (on an eight-point scale) in the Czech Republic and 6.96 points in the EU 28. The average length of the educational period for this occupational group was 17.1 years.

The European Commission has prepared a set of materials, the so-called circular economy package, which has triggered changes in Europe towards a circular economy. The whole economy will gradually adapt to these changes and will also be reflected in job changes and the need for new skills. Digitization will be one of the main drivers of the transition of institutions to the circular economy. Products and services will be more tailor-made. The principle lies in renting or sharing long-lasting products. One example of such a solution is a resource-sharing platform between institutions, where you can borrow not only medical equipment but also employees for a short time. As part of its Diamond Select Advance program, Philips rents an expensive magnetic resonance medical device. The hospital always has access to a state-of-the-art device that it would otherwise not afford, and Philips takes back the previous model to remanufacture. This approach requires close collaboration between hospitals, service organizations, refurbishment centers, and Philips. It is a long-term, sustainable win-win model (České sociální podnikání 2018).

Funding: This research was funded by Czech University of Life Sciences Prague Project IGA PEF, grant number 20181025.

Conflicts of Interest: The author declares no conflict of interest.

References

Artl, Josef, Markéta Artlová, and Eva Rublíková. 2002. *Analýza ekonomických časových řad s příklady*. Praha: Skripta VŠE.

Baumol, William. J., and William. G. Bowen. 1966. *Performing Arts: The Economic Dilemma*. Cambridge and Massachusetts: The MIT Press.

Baumol, William. J., Sue A. B. Blackman, and Edward N. Wolf. 1985. *Unbalanced Growth Revisited: Asymptotic Stagnacy and New Evidence*. Nashville: American Economic Association.

Becker, Gary S. 1964. *Human Capital*. Chicago: University of Chicago Press.
Bills, David B. 2004. *The Sociology of Education and Work*. Malden and Oxford: Blackwell Publishing.
Blažek, Jiří, and David Uhlíř. 2002. *Teorie regionálního rozvoje: nástin, kritika, klasifikace*. Praha: Karolinum.
Bočková, Věra. 2000. *Celoživotní vzdělávání—výzva nebo povinnost?* Olomouc: Univerzita Palackého.
Brown, Phillip, Anthony Hesketh, and Sara Williams. 2003. Employability in a Knowledge-Driven Economy. *Journal of Education and Work* 16: 107–26.
Bureš, Vladimír. 2007. *Znalostní management a proces jeho zavádění*. 1. vyd. Praha: Grada Publishing.
České sociální podnikání. 2018. Cirkulární ekonomika jako příležitost pro sociální podniky. Available online: https://ceske-socialni-podnikani.cz/socialni-podnikani/clanky/2893-cirkularni-ekonomika-jako-prilezitost-pro-socialni-podniky (accessed on 23 May 2018).
Český statistický úřad. 2018a. Česká republika v číslech. Available online: https://www.czso.cz/csu/czso/ceska-republika-od-roku-1989-v-cislech-2017-24bfnixod8#01 (accessed on 8 April 2018).
Český statistický úřad. 2018b. Klasifikace. Available online: https://www.czso.cz/csu/czso/klasifikace (accessed on 5 April 2018).
Český statistický úřad. 2018c. Národní účty. Available online: https://www.czso.cz/csu/czso/hdp_narodni_ucty (accessed on 21 April 2018).
Český statistický úřad. 2018d. Odvětví činnosti zaměstnaných. Available online: https://www.czso.cz/csu/czso/204r-k-odvetvi-cinnosti-zamestnanych-v-nh-m7e8nfaakv (accessed on 17 April 2018).
Český statistický úřad. 2018e. Zaměstnanost a nezaměstnanost. Available online: https://www.czso.cz/csu/czso/zamestnanost_nezamestnanost_prace (accessed on 8 April 2018).
Cipra, Tomáš. 1986. *Analýza časových řad s aplikacemi v ekonomii*. Praha: Státní nakladatelství technické literatury.
Cradle to Cradle. 2018. Product-Life. Available online: http://www.product-life.org/en/cradle-to-cradle (accessed on 13 April 2018).
Ellen MacArthur Foundation. 2012. Towards the Circular Economy: An Economic and Business Rationale for an Accelerated Transition. Available online: https://www.ellenmacarthurfoundation.org/assets/downloads/publications/Ellen-MacArthur-Foundation-Towards-the-Circular-Economy-vol.1.pdf (accessed on 23 April 2018).
Florida, Richard. 1995. Toward the learning region. *Futures* 27: 527–36. [CrossRef]
Govindan, Kannan, and Mia Hasanagic. 2018. A Systematic Review on Drivers, Barriers, and Practices towards Circular Economy: A Supply Chain Perspective. Available online: https://www.circularresourceslab.ch/wp-content/uploads/2018/08/Kannan-Govindan-Mia-Hasanagic-2018.pdf (accessed on 21 April 2018).
Hampl, Martin. 2001. *Regionální vývoj: Specifika české transformace, evropská integrace a obecná teorie*. Praha: PřF UK.
Hebák, Petr, Jiří Hustopecký, Eva Jarošová, and Iva Pecáková. 2007. *Vícerozměrné statistické metody (1)*. Praha: Informatorium.
Hirschman, Albert O. 1958. *The Strategy of Economic Development*. New Haven: Yale University Press.
Holman, Robert. 2001. *Dějiny ekonomického myšlení*. 2. vydání. Praha: C.H. Beck.
Howell, David. 2005. Chapter 2 Fighting Unemployment: The Limits of Free Market Orthodoxy. In *Wage Compression and the Unemployment Crisis: Labor Market Institutions, Skills, and Inequality-Unemployment Tradeoffs*. Edited by David R. Howell. New York: Oxford University Press.
Karpíšek, Zdeněk. 1999. *Fuzzy Reliability—Characteristics and Models*. Brno: VUT Brno.
Kirchherr, Julian, Denise Reike, and Marko Hekkert. 2017. Conceptualizing the Circular Economy: An Analysis of 114 Definitions. Available online: https://www.sciencedirect.com/science/article/pii/S0921344917302835 (accessed on 7 May 2018).
Korhonen, Jouni, Antero Honkasalo, and Jyri Seppälä. 2018. Circular Economy: The Concept and Its Limitations. Ecological Economics. Available online: https://www.researchgate.net/publication/318385030_Circular_Economy_The_Concept_and_its_Limitations (accessed on 17 May 2018).
Lundvall, Bengt A. 1992. *National Systems of Innovation; Toward a Theory of Innovation and Interactive Learning*. London: Pinter.
Maier, Gunther, and Franz Tödtling. 1998. *Regionálna a urbanistická ekonomika 2. Regionálny rozvoj a regionálna politika*. Bratislava: Elita.
Malinovský, Jan, and Jan Sucháček. 2006. *Velký anglicko—český slovník regionálního rozvoje a regionální politiky Evropské unie*. Ostrava: VŠB-Technická univerzita.
Mazouch, Petr, and Jakub Fisher. 2011. *Lidský kapitál měření, souvislosti, prognózy*. Praha: C.H. Beck.

Meloun, Milan, Jiří Militký, and Martin Hil. 2011. *Statistická analýza vícerozměrných dat v příkladech*. Praha: Academia Praha.

Myrdal, Gunnar. 1957. *Economic Theory and Under—Developed Regions*. London: Gerald Duckworks.

Národní ústav pro vzdělávání. 2018a. Absolventi, Charakteristiky a Perspektivy. Available online: http://www.infoabsolvent.cz/Temata/ClanekAbsolventi/4-4-03/Charakteristiky-a-perspektivy-profesnich-skupin-v-/34 (accessed on 3 May 2018).

Národní ústav pro vzdělávání. 2018b. Absolventi, podíly zaměstnaných podle profesních skupin. Available online: http://www.infoabsolvent.cz/Temata/ClanekAbsolventi/4-4-05/Podily-zamestnanych-podle-profesnich-skupin-ve-/34 (accessed on 3 May 2018).

Národní ústav pro vzdělávání. 2018c. Klasifikace. Available online: http://www.nuv.cz/nsk2/kvalifikace (accessed on 5 May 2018).

Národní vzdělávací fond. 2018. Vývoj v odvětvích. Available online: http://www.budoucnostprofesi.cz/cs/vyvoj-v-odvetvich.html (accessed on 12 May 2018).

Piore, Michael, and Charles F. Sabel. 1984. *The Second Industrial Divide: Possibilities for Prosperity*. New York: Basic Books.

Plesník, Vladimír. 2007. *Dlouhodobý aktivizační program pro nezaměstnané*. Krnov: REINTEGRA.

Redor, D. 1999. *Economie du travail et de l'emploi*. Paris: Eco-Montchrestien.

Samuelson, Paul, and William D. Nordhaus. 2008. *Ekonomie*. NS SVOBODA: Prague.

Saxenian, AnnaLee. 1994. *Regional Advantage: Culture and Competition in Silicon Valley and Route 128*. Cambridge: Harvard University Press.

Schultz, Theodore W. 1971. *Investment in Human Capital*. New York: The Free Press.

Severová, L. 2011. *Znalostní ekonomika a vzdělávání v mezinárodním kontextu*. Prague: Nakladatelství Alfa Publishing.

Trhlíková, Jana, Helena Úlovcová, and Jiří Vojtěch. 2006. *Sociální aspekty dlouhodobé nezaměstnanosti mladých lidí s nízkou úrovní vzdělání*. Praha: Národní ústav odborného vzdělávání.

Výzkumný ústav práce a sociálních věcí. 2018. Prognózování vzdělanostních potřeb na období 2008 až 2012—stav modelu a aktuální prognóza. Available online: http://praha.vupsv.cz/Fulltext/vz_292.pdf (accessed on 4 April 2018).

Zhijun, Feng, and Yan Nailing. 2007. Putting a Circular Economy into Practice in China. Available online: https://www.environmental-expert.com/Files/6063/articles/15086/art9.pdf (accessed on 11 April 2018).

© 2019 by the author. Licensee MDPI, Basel, Switzerland. This article is an open access article distributed under the terms and conditions of the Creative Commons Attribution (CC BY) license (http://creativecommons.org/licenses/by/4.0/).

Article

Waste Management Analysis in Developing Countries through Unsupervised Classification of Mixed Data

Giulia Caruso * and Stefano Antonio Gattone

Department of Philosophical, Pedagogical and Economic-Quantitative Sciences, University G. d'Annunzio Chieti-Pescara, Vle Pindaro n. 42, 65127 Pescara, Italy; gattone@unich.it
* Correspondence: giulia.caruso@unich.it

Received: 7 May 2019; Accepted: 6 June 2019; Published: 13 June 2019

Abstract: The increase in global population and the improvement of living standards in developing countries has resulted in higher solid waste generation. Solid waste management increasingly represents a challenge, but it might also be an opportunity for the municipal authorities of these countries. To this end, the awareness of a variety of factors related to waste management and an efficacious in-depth analysis of them might prove to be particularly significant. For this purpose, and since data are both qualitative and quantitative, a cluster analysis specific for mixed data has been implemented on the dataset. The analysis allows us to distinguish two well-defined groups. The first one is poorer, less developed, and urbanized, with a consequent lower life expectancy of inhabitants. Consequently, it registers lower waste generation and lower CO_2 emissions. Surprisingly, it is more engaged in recycling and in awareness campaigns related to it. Since the cluster discrimination between the two groups is well defined, the second cluster registers the opposite tendency for all the analyzed variables. In conclusion, this kind of analysis offers a potential pathway for academics to work with policy-makers in moving toward the realization of waste management policies tailored to the local context.

Keywords: cluster analysis; unsupervised classification; mixed data; circular economy; waste management

1. Introduction

In recent years, the term circular economy has gained much attention. It refers to a system of production and consumption providing minimal losses of materials and energy through extensive reuse, recycling, and recovery Haupt et al. (2017). In other words, it is an economic system for which is essential to recycle materials from waste in order "to close the cycle".

Furthermore, the increase in global population and the improvement of living standards in developing countries has resulted in higher solid waste generation in the areas under investigation Abarca-Guerrero et al. (2013); Minghua et al. (2009). Consequently, nowadays waste management is at the center of a very lively debate Fernández-González et al. (2017); Schneider et al. (2017).

Waste disposal should be gradually eliminated, and where this is not possible, it should be monitored in order to be safe for human health and the environment. Consequently, solid waste management increasingly represents a challenge since municipalities must provide an efficient system to their population. However, they often have to struggle with complexity and with lack of both organization and financial resources Burntley (2007).

Among other things, sustainable waste management is able to reduce the incidence of health problems, the emission of greenhouse gases, and the deterioration of landscape, water, and air caused by landfilling Cucchiella et al. (2017). However, it not only represents a contribution to environmental protection, but it also pays off economically Nelles et al. (2016): waste management practices, indeed, can be cost-saving and generate revenue opportunities Romero-Hernández and Romero (2018).

In fact, waste can be a useful source of raw materials and energy, too. Metals, glass, and textiles have long been collected and put to new use; for example, the extraction of nickel and cobalt from raw materials, as well as from waste, is strategically important for industry and society Komnitsas et al. (2019). Waste can be turned into energy too, enabling the value of products, materials, and resources to be maintained on the market for as long as possible, minimizing waste and resource use in the wake of the objectives of circular economy Malinauskaite et al. (2017).

For all these reasons, and since the circular economy is an important issue for the future and the competitiveness of businesses Garcia-Muiña et al. (2018), solid waste management might represent an opportunity for the municipal authorities of developing countries, especially those characterized by a low-income Minghua et al. (2009).

In conclusion, in a circular economy wastes are considered a resource, especially from an economic point of view, consequently attracting an increasing number of industrial actors, policy-makers, and researchers Cucchiella et al. (2017). Thus, in the last years, a large number of studies have tried to detect factors influencing waste management systems in developing countries.

Therefore, since the awareness of a variety of factors related to waste management might prove to be particularly significant Ghinea and Gavrilescu (2019); Zeller et al. (2019); Zohoori and Ghani (2017), this paper aims to analyze some of them in the developing countries involved in the study.

From our analysis, it may be concluded that a study of this kind offers a potential pathway for academics to work with policy-makers in moving toward the realization of waste management policies tailored to the local context.

2. Materials and Methods

Real data often consists of mixed variables, that is, both continuous and categorical ones; an example is provided by the dataset analyzed in this paper, produced by Abarca-Guerrero (2014), the variables of which are described in Table 1. Traditionally, cluster analysis has only focused on datasets composed of a single type of variable (all quantitative or all qualitative). For this reason, researchers dealing with mixed data usually convert them into a single data type, transform the categorical variables into binary ones and consequently apply methods for numeric variables, or transform continuous variables into categorical ones Dougherty et al. (1995); Ichino and Yaguchi (1994).

Indeed, clustering methods specific to mixed data are less encountered in the literature.
Some traditional methods are:

- data pre-processing, that is, all variables are converted to the same scale, either numerical to categorical or vice-versa;
- distance measures specifically developed for mixed datasets.

With regards to data pre-processing, these algorithms are essentially created for purely categorical attributes, although they have also been applied to mixed data after a transformation of numerical attributes to categorical ones (discretization). In general, these kinds of algorithms can be applied to mixed data through a discretization process that may, nevertheless, produce a loss of important information Caruso et al. (2018).

One example is represented by the dummy coding of all categorical variables. But this increases the dataset's dimensionality, representing a problem when the number of categorical variables simultaneously increases with the size of the data. Another disadvantage is that any semantic similarity in the original dataset is lost in the transformed one. Finally, coding strategies imply a difficult choice of weights representing categorical attributes Foss et al. (2016).

An alternative to recoding categorical or continuous variables is to use a dissimilarity measure, taking into account the different types of data Caruso (2019). A common approach is to use the Gower distance Gower (1971).

2.1. Clustering Mixed Data

Let $X = \{x_1, x_2, \ldots, x_n\}$ denote a set of n objects and $x_i = [x_{i1}, x_{i2}, \ldots, x_{iL}]$ indicate an object constituted by L variables. Since the L variables of the considered dataset are both continuous and categorical, it is possible to write $L = Q + C$, where Q corresponds to the number of numeric variables and C to the number of categorical ones. $\mathcal{C} = \{l_1^C, \ldots l_C^C\}$ is a subset identifying the qualitative variables and $\mathcal{Q} = \{l_1^Q, \ldots l_Q^Q\}$ is a subset denoting the quantitative ones. The aim of clustering is to assign the n objects contained in X to K separate clusters. When clustering mixed datasets, the main problem is to determine how close or how far apart objects are from each other.

There are different approaches presenting different ways to combine distance measures for numerical variables and distance measures for categorical ones into a single cost function Caruso (2019); Caruso et al. (2019, in printb); Everitt (1974); Huang (1997).

2.2. The Huang Method

Huang (1998) presented a so-called K-prototypes algorithm, which is based on the K-means method but overcomes its quantitative data limitation, preserving, at the same time, its efficiency. The algorithm groups the objects in clusters against k prototypes. The updates occur in a dynamical manner, so as to minimize the following objective function:

$$E = \sum_{k=1}^{K}\sum_{i=1}^{n} u_{ik}\Phi_H(x_i, V_k), \qquad (1)$$

where u_{ik} is an element of a partition matrix $U_{n\times k}$, and $\Phi_H(x_i, V_k)$ is a dissimilarity measure for mixed data between the objects x_i and V_k.

$V_k = [v_{k1}, v_{k2}, \ldots, v_{kL}]$ is the prototype or representative vector for cluster k. U represents a hard partition matrix, where $u_{ik} \in \{0,1\}$, and $u_{ik} = 1$ if x_i is allocated to cluster k.

The Huang dissimilarity measure for mixed data is defined as

$$\Phi_H(x_i, V_k) = \sum_{l \in \mathcal{Q}}(x_{il} - v_{kl})^2 + \gamma_k \sum_{l \in \mathcal{C}} \delta(x_{il}, v_{kl}), \qquad (2)$$

where the first term is the squared Euclidean distance, whereas the second one is defined as $\delta(r,t) = 0$ for $r = t$ and $\delta(r,t) = 1$ for $r \neq t$. γ_k is a weight for categorical variables in cluster k.

The internal term in Equation (1) can be defined as $E_k = \sum_{i=1}^{n} u_{ik} S(x_i, V_k)$. It measures the total dissimilarity of objects in cluster k from their prototype V_k. The quantity E_k could be considered as the total cost of allocating the objects $x_i (i \in C_k)$ to cluster k.

This term may be rewritten as

$$\begin{aligned} E_k &= \sum_{i=1}^{n} u_{ik} \sum_{l \in \mathcal{Q}}(x_{il} - v_{kl})^2 + \gamma_k \sum_{i=1}^{n} u_{ik} \sum_{l \in \mathcal{C}} \delta(x_{il}, v_{kl}) \\ &= E_k^{\mathcal{Q}} + \gamma_k E_k^{\mathcal{C}}, \end{aligned} \qquad (3)$$

where $E_k^{\mathcal{Q}}$ and $E_k^{\mathcal{C}}$ represent the dissimilarity of the objects in cluster k for the quantitative and the qualitative variables, respectively. In order to minimize these two components, let $V_k^{\mathcal{Q}}$ and $V_k^{\mathcal{C}}$ be the prototypes for cluster k for the numerical and categorical variables, respectively.

$E_k^{\mathcal{Q}}$ is minimized with the usual update of the K-means algorithm for continuous variables. That is, the generic component of $V_k^{\mathcal{Q}}$ is the arithmetic mean:

$$v_{kl} = \frac{1}{n_k}\sum_{i=1}^{n} u_{ik} x_{il} \qquad l \in \mathcal{C}, \qquad (4)$$

where n_k is the number of objects in cluster k. Let $\mathcal{W}_l = \{w_{l,1}, w_{l,2}, \ldots, w_{l,m_l}\}$ be the set enclosing the distinct values of the l-th categorical variable, and let $p_l(w_{l,j}|k)$ be the probability that value $w_{l,j}$ is observed in cluster k.

It is possible to rewrite $E_k^\mathcal{C}$ in (3) as

$$E_k^\mathcal{C} = \sum_{l \in \mathcal{C}} n_k \left[1 - p(v_{kl} \in \mathcal{W}_l|k)\right] . \tag{5}$$

In Equation (5), $E_k^\mathcal{C}$ is minimized by selecting the categorical values of the prototype $W_k^\mathcal{C}$, such that $p(v_{kl} \in \mathcal{W}_l|k) \geq p(w_{l_j} \in \mathcal{W}_l|k)$ for $v_{kl} \neq w_{l,j}$ for all categorical variables.

On the basis of the Huang algorithm, by minimizing (1), we implemented a cluster analysis with a number of clusters equal to $K = 2$. This choice was made based on the Silhouette index Rousseeuw (1987); since higher values corresponds to better results, the resultant (optimal) maximum value precisely corresponds to 2.

3. Results

3.1. The Dataset

The dataset Abarca-Guerrero (2014) used in this application has been extracted from the data archive of the "4TU.Centre for Research Data" in the Netherlands, and it regards the period of 1985–2011. It contains information on factors influencing the municipal waste management system in 22 developing countries; each of these is associated with more than one observation.

The dataset considers some key factors affecting waste management systems, in particular the country performance in terms of public health (life expectancy at birth), economy (gross domestic product/capita/year), and environment (CO_2-emissions/capita).

Other general parameters characterizing the countries are the urban population, the kind of climate, and precipitation. Furthermore, waste-specific parameters have been considered: the waste generation rate (kg/capita/day) and the sophistication of waste collection. The latter can be articulated as 1 = no organized collection of solid waste; 2 = collection based on manpower only; 3 = collection based on both manpower and draught animal; 4 = collection based on motorized transport but no compactor used; and 5 = collection based on motorized transport and compactor used. Other parameters include the existence of a recycling culture and the presence of municipality awareness campaigns, of recyclable-material-buying companies, and of recycling companies, the latter two specifically in the surroundings of the city. In conclusion, the analyzed data consist of a selection of 11 variables (6 continuous and 5 categorical), described in detail in Table 1, for a total of 50 observations Abarca-Guerrero (2014).

Table 1. Description of all of the analyzed variables.

Variable	Type	Description
Urban population	Continuous	% of urban population
Waste generation	Continuous	Waste generation rate (kg/capita/day)
CO_2	Continuous	CO_2-emission/capita in percentage of disposable income
GDP	Continuous	Gross domestic product/capita/year
Life Expectancy	Continuous	Life expectancy at birth (years)
Municipality campaigns	Continuous	Recycling awareness campaigns supported by the municipality: 1 = yes 2 = no

Table 1. Cont.

Variable	Type	Description
Waste collection combination	Categorical	Waste collection combination: 1 = no collection 2 = animal power 3 = man power 4 = animal+man 5 = mechanized
Climate	Categorical	Climate: 1 = equatorial 2 = arid 3 = warm temperature 4 = snow
Precipitation	Categorical	Precipitation: 1 = desert 2 = steppe 3 = fully humid 4 = summer dry 5 = winter dry 6 = monsoonal
Recyclable-material-buying companies	Categorical	Companies buying recyclable materials in the surroundings of the city: 1 = none 2 = few 3 = some 4 = many 5 = very many
Recycling companies	Categorical	Recycling companies in the surroundings of the city: 1 = none 2 = few 3 = some 4 = many 5 = very many

3.1.1. Internal Indexes

Since the ground truth (i.e., an empirical evidence) Han et al. (2011) is not given for this dataset, it is not possible to compute the external indexes. The internal ones are shown below.

We compared several methods for clustering mixed data types, namely those of: Huang (1997), Ahmad and Dey (2007), and Cheung and Jia (2013).

The relevant validity of cluster results was evaluated through the above-mentioned indexes and the Huang method proved to be the one yielding the best results, namely the highest values of the Calinski–Harabasz index (CH) Calinski and Harabasz (1974) and of the Silhouette index (SHI) Rousseeuw (1987), both computed on quantitative variables. The results provided by these methods are shown in Table 2.

Table 2. Internal indexes. CH—Calinski–Harabasz index; SHI—Silhouette index.

Method	CH	SHI
Huang	13.23	0.21
Ahmad & Dey	10.15	0.209
Cheung & Jia	10.68	0.189

3.1.2. Analysis of Quantitative Variables

First of all, we provide the descriptive analysis of the quantitative variables used: the mean, the standard deviation, and the values corresponding to the 1st, the 2nd, and the 3rd quartiles, as shown in Table 3.

Table 4 displays, for each cluster, the mean value of the analyzed quantitative attributes. With regards to the variable "waste generation", the two groups have a weak cluster structure, that is clusters values are very similar between them, whereas they have a strongest structure with regards to the variable "GDP". With regards to the "percentage of urban population", the relevant overall mean equals 51.12. Thus, the first cluster mean is lower than the average one, whereas the second one is higher. For what concerns the "waste generation rate", instead, the overall mean corresponds to 0.61. In this case as well, the first cluster mean is lower than the overall one, whereas the second one is higher. However, in this case the separation between clusters is less marked. With regards to the "CO_2 emissions", the overall mean value is 2.28, so the first cluster mean is lower than the overall one, whereas the second one is higher. For what concerns the "GDP", the overall mean value is equal to 3825, so the first cluster mean is lower than the average, whereas the second one is significantly higher. With regards to "life expectancy", the overall mean equals 68.46; in this case as well, the first mean is lower and the second one is higher. In summary, for all of the analyzed quantitative variables, the first cluster is below the overall mean whereas the second cluster overcomes it.

Table 3. Descriptive statistics of quantitative variables.

Variables	Mean	Standard Deviation	Q_1	Q_2	Q_3
%UrbPop	51.12	19.35	33.50	57	65.75
WasGen	0.61	0.28	0.41	0.50	0.82
CO2	2.28	2.51	0.80	1.40	3.50
GDP	3825	6747.90	1069	2349	4469
LifeExp	68.46	8.27	66	71	73

Table 4. Mean values of quantitative variables for each cluster.

Cluster	Size	%UrbPop	WasGen	CO2	GDP	LifeExp
1	29	42.24	0.57	1.72	2096.38	66.24
2	21	63.38	0.68	3.06	6212.29	71.52

3.1.3. Analysis of Qualitative Variables

Table 5 shows the overall distribution of the variable "municipality campaigns". The prevailing modality is represented by the presence of recycling awareness campaigns supported by the municipality. Table 6, instead, shows more in detail the distribution of the categorical variable "municipality campaigns" for each of the two clusters. The first one, which has the strongest cluster structure, is characterized by the presence of recycling awareness campaigns supported by the municipality, whereas the second cluster is characterized by their absence.

Table 5. Overall distribution of the categorical variable "municipality campaigns".

Municipality Campaigns	Distribution
Yes	0.68
No	0.32

Table 6. Distribution of the categorical variable "municipality campaigns".

Municipality Campaigns	Clusters	
	1	2
Yes	0.90	0.38
No	0.10	0.62

Table 7 shows the overall distribution of the variable "waste collection combination". The prevailing modality is represented by "animal power", followed by "mechanized methods".

In Table 8, the two clusters are clearly outlined. In the first, the prevailing modality is represented by "animal power", whereas in the second one it corresponds to "mechanized" methods. By comparing the obtained clusters with the relevant overall distribution, the first cluster registers a higher use of methods based on "animal power" than in the overall distribution. The same applies to the modality "mechanized" methods in the second cluster.

Table 7. Overall distribution of the categorical variable "waste collection combination".

Waste Collection Combination	Distribution
No collection	0.02
Animal power	**0.42**
Man power	0.12
Animal + man	0.12
Mechanized	**0.32**

Table 8. Distribution of the categorical variable "waste collection combination".

Waste Collection Combination	Clusters	
	1	2
No collection	0.00	0.05
Animal power	**0.66**	0.10
Man power	0.17	0.05
Animal + man	0.00	0.29
Mechanized	0.17	**0.52**

Table 9 shows the overall distribution of the variable "climate". The modality with the overwhelming majority is "equatorial".

Table 10 shows that in both clusters, the prevailing modality is represented by the modality "equatorial", in the wake of the overall distribution.

Table 9. Overall distribution of the categorical variable "climate".

Climate	Distribution
Arid	0.18
Equatorial	**0.72**
Snow	0.02
Warm	0.08

Table 10. Distribution of the categorical variable "climate".

Climate	Clusters	
	1	2
Arid	0.17	0.19
Equatorial	**0.76**	**0.67**
Snow	0.03	0.00
Warm	0.03	0.14

Table 11, instead, shows the overall distribution of the variable "precipitation". The prevailing modality is represented by "fully humid".

In Table 12, the distribution of the two clusters is shown. The first cluster is characterized by the prevalence of the modality "monsoonal", whereas in the second cluster the most frequent modality is "fully humid", exactly like in the overall distribution.

Table 11. Overall distribution of the categorical variable "precipitation".

Precipitation	Distribution
Desert	0.06
Fully humid	**0.36**
Monsoonal	0.24
Steppe	0.08
Summer dry	0.06
Winter dry	0.20

Table 12. Distribution of the categorical variable "precipitation".

Precipitation	Clusters	
	1	2
Desert	0.10	0.00
Fully humid	0.21	**0.57**
Monsoonal	**0.41**	0.00
Steppe	0.07	0.10
Summer dry	0.07	0.05
Winter dry	0.14	0.29

Table 13 shows the overall distribution of the variable "recyclable-material-buying companies". The prevailing modality is represented by "some".

In Table 14, the distribution of the two clusters is shown. The first cluster is characterized by a prevalence of the modality "some", exactly as in the overall distribution, whereas in the second cluster the most frequent modality is "none".

Table 15 shows the overall distribution of the variable "recycling companies in the surroundings of the city". The prevailing modality is represented by "none".

In Table 16, the distribution of the two clusters is shown. The first cluster has a prevalence of the modality "some", whereas in the second cluster the most frequent modality is "none", exactly as in the overall distribution.

Table 13. Overall distribution of the categorical variable "recyclable-material-buying companies".

Recyclable-Material-Buying Companies	Distribution
none	0.28
few	0.18
some	0.38
many	0.16
very many	0.00

Table 14. Distribution of the categorical variable "recyclable-material-buying companies".

Recyclable-Material-Buying Companies	Clusters	
	1	2
none	0.00	0.67
few	0.10	0.29
some	0.62	0.05
many	0.28	0.00
very many	0.00	0.00

Table 15. Overall distribution of the categorical variable "recycling companies in the surroundings of the city".

Recycling Companies	Distribution
none	0.34
few	0.30
some	0.24
many	0.10
very many	0.02

Table 16. Distribution of the categorical variable "recycling companies in the surroundings of the city".

Recycling Companies	Clusters	
	1	2
none	0.14	0.62
few	0.31	0.29
some	0.38	0.05
many	0.17	0.00
very many	0.00	0.00

4. Discussion

On the basis of our analysis of quantitative variables, it appears that the first cluster is characterized by a lower percentage of urban population, lower levels of GDP, and a lower life expectancy. As a consequence of limited urbanization and greater poverty, this group registers lower rates of waste generation and of CO_2 emissions. Since the cluster discrimination between the two groups is well defined, the second cluster registers the opposite tendency for all of the above-mentioned variables, namely higher levels of GDP and a stronger percentage of urban population, with a consequently higher life expectancy. The higher urbanization corresponds to higher levels of waste generation and of CO_2 emissions. In more detail, in order to better describe the distribution of the quantitative variables analyzed, each of these has been represented through a box plot Cleveland (1993).

With regards to the percentage of urban population (Figure 1), the overall median of the distribution equals 57.00, whereas the overall mean is 51.12; the mean of the first cluster is lower than this, whereas the one associated with the second cluster is higher. In the first cluster, the interquartile distance is much higher than in the second one, denoting a greater dispersion of the 50% most central observations around the median. On the other hand, since the interquartile distance of the second cluster is lower, the 50% most central observations are highly concentrated around the median. Furthermore, since in the first cluster the distances between each quartile and the median are quite different from one another, the distribution is asymmetric. In the second cluster, instead, the distances are more similar between these, denoting a lower asymmetry of the distribution.

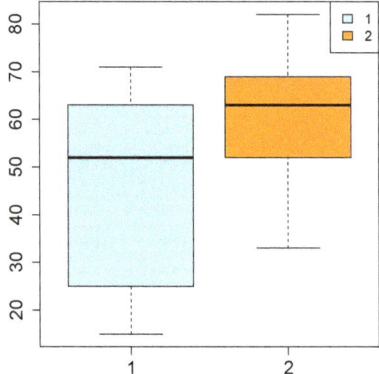

Figure 1. Boxplot of the variable "urban population" in clusters 1 and 2.

For what concerns the waste generation (Figure 2), the overall median of the distribution equals 0.50, whereas the overall mean is 0.61; the mean of the first cluster is lower than this, whereas the one associated with the second cluster is higher. In the first cluster, the interquartile distance is lower than in the second one. Thus, in the first group there is a low dispersion of the 50% most central observations around the median. On the other hand, since the interquartile distance of the second cluster is higher, the 50% most central observations are less concentrated around the median. Furthermore, in the first cluster the two distances are quite different from one another, denoting a very asymmetric distribution, whereas in the second group these distances are more similar, resulting in a slightly lower asymmetry of the distribution.

With regards to the CO_2 emissions (Figure 3), the overall median of the distribution equals 1.40, whereas the overall mean is 2.28; the mean of the first cluster is lower than this, whereas the one associated to the second cluster is higher. In the first cluster, the interquartile distance is lower than in the second one. Thus in the first group, the dispersion of the 50% most central observations around the median is lower. On the other hand, since the interquartile distance of the second cluster is higher, the 50% most central observations are less concentrated around the median. Furthermore, in the first cluster the two distances are very similar to one another, denoting a symmetric distribution, whereas in the second cluster they are less similar, indicating the asymmetry of the distribution.

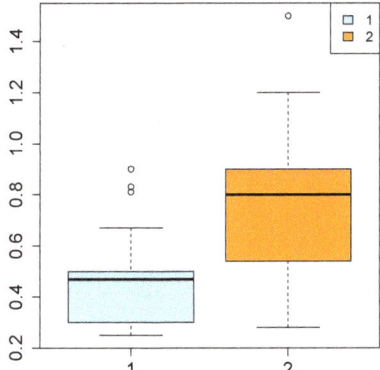

Figure 2. Boxplot of the variable "waste generation" in clusters 1 and 2.

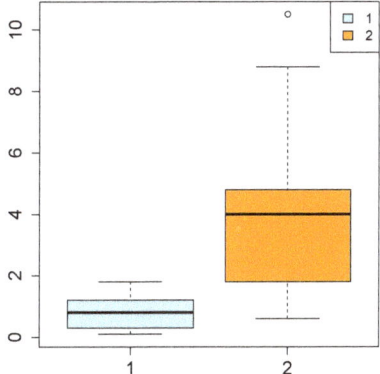

Figure 3. Boxplot of the variable "CO_2" in clusters 1 and 2.

With regards to the GDP (Figure 4), the overall median of the distribution equals 2349, whereas the mean is 3825; the median of the first cluster is lower than this, whereas the one associated with the second cluster is higher. In the first cluster, the interquartile distance is lower than in the second one. Thus, in the first group the dispersion of the the 50% most central observations around the median is lower. Since the interquartile distance of the second cluster is higher, instead, the 50% most central observations are less concentrated around the median. Furthermore, in the first cluster the two interquartile distances are quite different from one another, so the distribution is asymmetric. In the second cluster, instead, they are more similar, denoting the lower asymmetry of the distribution.

For what concerns life expectancy (Figure 5), the overall median of the distribution equals 71.00, whereas the overall mean is 68.46, thus the mean of the first cluster is lower than this, whereas the one associated with the second cluster is higher. In the first group, the interquartile distance is higher than in the second one. Thus, in the first group the dispersion of the 50% most central observations around the median is higher. On the other hand, since the interquartile distance of the second cluster is lower, the 50% most central observations are highly concentrated around the median. Furthermore, in the first cluster the two distances are quite different from one another, so the distribution is asymmetric, whereas in the second group the distances are more similar, denoting the lower asymmetry of the distribution.

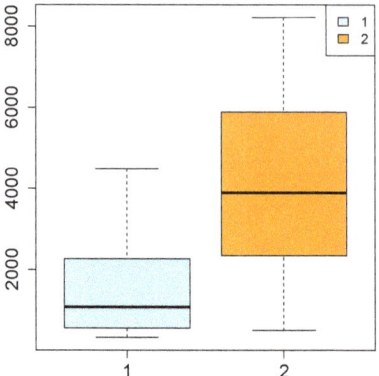

Figure 4. Boxplot of the variable "GDP" in clusters 1 and 2.

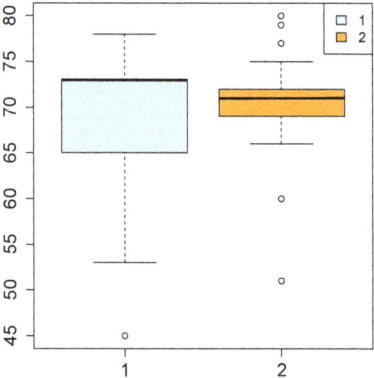

Figure 5. Boxplot of the variable "life expectancy" in clusters 1 and 2.

With regards to qualitative variables, instead, the first cluster is characterized by the overwhelming majority of recycling awareness campaigns supported by the municipality, the waste is mainly collected through animal power, and there are some recyclable-material-buying companies and some recycling companies in the surrounding areas of the cities. The countries falling under this category are mostly characterized by monsoonal precipitation and are the following: Ethiopia, Sri Lanka, Thailand, China, Peru, Tanzania, India, Bangladesh, Nepal, Malawi, Zambia, Nicaragua, Kenya, and the Philippines.

The second cluster, instead, is mainly characterized by the absence of recycling awareness campaigns supported by the municipality, the waste is mainly collected through mechanized tools, but it is mostly characterized by the absence of recyclable-material-buying companies and of recycling companies in the surrounding areas of the cities. Furthermore, it is characterized by the prevalence of a fully humid climate. The countries falling into this cluster are Turkey, Suriname, Costa Rica, Ecuador, Pakistan, and Bhutan, whereas Indonesia and South Africa are in the overlapping area of the two clusters.

5. Conclusions

Since nowadays more and more applications are based on datasets composed of mixed data, there is an ever-growing interest in cluster analysis. Due to their characteristics, traditional methods are unable to capture, store, manage, and analyze these datasets. A cluster analysis implemented on such a dataset has huge potential; however, most clustering algorithms are designed to exclusively handle one type of data at a time, being unable to analyze mixed data simultaneously. The use of cluster analysis for mixed data represents an element of innovation, especially in the waste management sector, since until now only the traditional cluster analysis has been applied in this framework.

Certainly, the research in this area is far from being complete. There are quite a few methods in the literature, but further advancements in this field are needed Caruso (2019). Furthermore, in the wake of this work, future research will be focused on the development of new cluster analysis techniques for mixed data and on the consequent creation of dedicated software packages, also with the aim of widening the number of potential users of this method Caruso et al. (2019).

The basis for future developments will take into consideration the results yielded from the applications described in Section 3 and from an interesting insight provided by the work of Diday and Govaert Diday and Govaert (1977). They propose an adaptive clustering that consists in a dynamic procedure and is useful for calibrating the weights of variables used in the clustering.

Usually, indeed, all of the variables participate in the cluster analysis with the same importance, but since some of them may be more discriminant than others, or better characterize a cluster, there are some ways to correctly consider their different values Irpino et al. (2016).

One strategy consists in assigning a weight to each variable in advance, on the basis of a prior knowledge, and then performing a cluster analysis; a future development could consist in computing the weights for each variable in an automatic way Caruso (2019). In this context, Diday and Govert Diday and Govaert (1977) proposed using an adaptive distance when clustering real data. It is necessary to introduce a weighting step in the optimization process, generating a set of weights; each of these corresponds to a variable and measures its importance in the cluster analysis. While Diday and Govert's proposal is only focused on quantitative variables, a further advancement could be to extend it to both quantitative and qualitative data.

Furthermore, an additional input for our future research could be to extend this kind of analysis to our recent study Caruso et al. (in printa). Moreover, since clustering is also at the center of a very lively debate Di Battista et al. (2016), Di Battista and Fortuna (2016), Fortuna and Maturo (2018), Fortuna et al. (2018) in the functional framework, a further and interesting possible development could be to also consider this kind of approach in our future research.

Author Contributions: Conceptualization G.C.; methodology G.C.; software, G.C.; formal analysis, G.C.; data curation, G.C.; supervision, S.A.G.

Funding: This research received no external funding.

Conflicts of Interest: The authors declare no conflict of interest.

References

Abarca-Guerrero, Lilliana. 2014. Municipal Waste Management Data Set. Eindhoven University of Technology. Available online: https://doi.org/10.4121/uuid:31d9e6b3-77e4-4a4c-835e-5c3b211edcfc (accessed on 6 June 2019).

Abarca-Guerrero, Lilliana, Ger Maas, and William Hogland. 2013. Solid waste management challenges for cities in developing countries. *Waste Management* 33: 220–32. [CrossRef] [PubMed]

Ahmad, Amir, and Lipika Dey. 2007. A k-mean clustering algorithm for mixed numeric and categorical data. *Data & Knowledge Engineering* 63: 503–27.

Burntley, Stephen. 2007. A review of municipal solid waste composition in the United Kingdom. *Waste Management Journal* 27: 1274–85. [CrossRef] [PubMed]

Calinski, Tadeusz, and Joachim Harabasz. 1974. A dendrite method for cluster analysis. *Communications in Statistics* 3: 1–27.

Caruso, Giulia. 2019. Cluster Analysis for Mixed Data. Ph.D. Dissertation, University G. d'Annunzio Chieti-Pescara, Chieti, Italy.

Caruso, Giulia, Stefano Antonio Gattone, Francesca Fortuna, and Tonio Di Battista. 2018. Cluster Analysis as a Decision-Making Tool: A Methodological Review. In *Decision Economics: In the Tradition of Herbert A. Simon's Heritage*. Advances in Intelligent Systems and Computing. Edited by Edgardo Bucciarelli, Shu-Heng Chen and Juan Corchado. Berlin: Springer International Publishing, Vol. 618, pp. 48–55.

Caruso, Giulia, Stefano Antonio Gattone, Antonio Balzanella, and Tonio Di Battista. 2019. Cluster analysis: An application to a real mixed-type data set. In *Models and Theories in Social Systems*. Studies in Systems, Decision and Control. Edited by Cristina Flaut, Sarka Hoskova-Mayerova and Daniel Flaut. Berlin: Springer International Publishing, Vol. 179, pp. 525–33.

Caruso, Giulia, Tonio Di Battista, and Stefano Antonio Gattone. In printa. A micro-level analysis of regional economic activity through a PCA approach. In *Decisions Economics: Complexity of Decisions and Decisions for Complexity*. Advances in Intelligent Systems and Computing. Edited by Edgardo Bucciarelli, Shu-Heng Chen and Juan Corchado. Berlin: Springer International Publishing.

Caruso, Giulia, Stefano Antonio Gattone, Francesca Fortuna, and Tonio Di Battista. In printb. Cluster Analysis for mixed data: An application to credit risk evaluation. In *Book of Short Papers IES 2019*. Advances in Intelligent Systems and Computing. Edited by Matilde Bini, Pietro Amenta, Antonello D'Ambra and Ida Camminatiello. Naples: Cuzzolin.

Cheung, Yiu-ming, and Hong Jia. 2013. Categorical-and-numerical-attribute data clustering based on a unified similarity metric without knowing cluster number. *Pattern Recognition* 46: 2228–38. [CrossRef]

Cleveland, William. 1993. *Visualizing Data*. New York: Murray Hill.

Cucchiella, Federica, Idiano D'Adamo, Massimo Gastaldi, Lenny Koh, and Paolo Rosa. 2017. A comparison of environmental and energetic performance of European countries: A sustainability index. *Renewable and Sustainable Energy Reviews* 78: 401–13. [CrossRef]

Di Battista, Tonio, and Francesca Fortuna. 2016. Clustering dichotomously scored items through functional data analysis. *Electronic Journal of Applied Statistical Analysis* 9: 433–50.

Di Battista, Tonio, Angela De Sanctis, and Francesca Fortuna. 2016. Clustering functional data on convex function spaces. In *Studies in Theoretical and Applied Statistics, Selected Papers of the Statistical Societies*. Berlin: Springer, pp. 105–14.

Diday, Edwin, and Gerard Govaert. 1977. Classification Automatique avec Distances Adaptatives. *R.A.I.R.O. Informatique Computer Science* 11: 329–49.

Dougherty, James, Ron Kohavi, and Mehran Sahami. 1995. Supervised and unsupervised discretization of continuous features. In *Machine Learning: Proceedings of the Twelfth International Conference, Tahoe City, CA, USA, July 9–12*. San Francisco: Morgan Kaufmann Publishers, pp. 194–202.

Everitt, Brian. 1974. *Cluster Analysis*. London: Heinemann Educational Books Ltd.

Fernández-Gonzalez, Jose-Manuel, Alejandro Luis Grindlay, Francisco Serrano-Bernardo, Maria Isabel Rodríguez-Rojas, and Montserrat Zamorano. 2017. Economic and environmental review of Waste-to-Energy systems for municipal solid waste management in medium and small municipalities. *Waste Management* 67: 360–74. [CrossRef]

Fortuna, Francesca, and Fabrizio Maturo. 2018. K-means clustering of item characteristic curves and item information curves via functional principal component analysis. *Quality and Quantity* [CrossRef]

Fortuna, Francesca, Fabrizio Maturo, and Tonio Di Battista. 2018. Clustering functional data streams: Unsupervised classification of soccer top players based on Google trends. *Quality and Reliability Engineering International* 34: 1448–60. [CrossRef]

Foss, Alex, Marianthi Markatou, Bonnie Ray, and Aliza Heching. 2016. A semiparametric method for clustering mixed data. *Machine Learning* 105: 419–58. [CrossRef]

Garcia-Muiña, Fernando, Rocio González-Sánchez, Anna Maria Ferrari, and Davide Settembre-Blundo. 2018. The Paradigms of Industry 4.0 and Circular Economy as Enabling Drivers for the Competitiveness of Businesses and Territories: The Case of an Italian Ceramic Tiles Manufacturing Company. *Social Sciences* 7: 255. [CrossRef]

Ghinea, Cristina, and Maria Gavrilescu. 2019. Solid Waste Management for Circular Economy: Challenges and Opportunities in Romania—The Case Study of Iasi County. In *Towards Zero Waste*. Greening of Industry Networks Studies. Edited by Maria-Laura Franco-García, Jorge Carlos Carpio-Aguilar and Hans Bressers. Berlin: Springer International Publishing, vol. 6, pp. 25–60.

Gower, John. 1971. A general coefficient of similarity and some of its properties. *Biometrics* 27: 857–71. [CrossRef]

Han, Jiawei, Micheline Kamber, and Jian Pei. 2011. *Data Mining: Concepts and Techniques*, 3rd ed. Burlington: Morgan Kaufmann.

Haupt, Melanie, Carl Vadenbo, and Stefanie Hellweg. 2017. Do we have the right performance indicators for the circular economy?: Insight into the Swiss waste management system. *Journal of Industrial Ecology* 21: 615–27. [CrossRef]

Huang, Zhexue. 1997. Clustering large data sets with mixed numeric and categorical values. Paper presented at the First Pacific-Asia Conference on Knowledge Discovery and Data Mining, Singapore, February 23–24; pp. 21–34.

Huang, Zhexue. 1998. Extensions to the k-means algorithm for clustering large data sets with categorical values. *Data Mining and Knowledge Discovery* 2: 283–304. [CrossRef]

Ichino, Manabu, and Hirotake Yaguchi. 1994. Generalized Minkowski metrics for mixed feature type data analysis. *IEEE Transactions on Systems, Man and Cybernetics* 24: 698–708. [CrossRef]

Irpino, Antonio, Rosanna Verde, and Francisco De Carvalho. 2016. Fuzzy clustering of distribution-valued data using adaptive L2 Wasserstein distances. *arXiv*: ArXiv:1605.00513

Komnitsas, Kostas, Evangelos Petrakis, Georgios Bartzas, and Vassiliki Karmali. 2019. Column leaching of low-grade saprolitic laterites and valorization of leaching residues. *Science of The Total Environment* 665: 347–57. [CrossRef]

Malinauskaite, Jurgita, Hussam Jouhara, Dina Czajczyńska, Peter Stanchev, Evina Katsou, Pawel Rostkowski, and Lorna Anguilano. 2017. Municipal solid waste management and waste-to-energy in the context of a circular economy and energy recycling in Europe. *Energy* 141: 2013–44. [CrossRef]

Minghua, Zhu, Xiumin Fan, Alberto Rovetta, Qichang He, Federico Vicentini, Bingkai Liu, Alessandro Giusti, and Yi Liu. 2009. Municipal solid waste management in Pudong New Area, China. *Waste Management Journal* 29: 1227–33. [CrossRef] [PubMed]

Nelles, Michael, Jennifer Gruenes, and Gert Morscheck. 2016. Waste management in Germany-development to a sustainable circular economy? In *Procedia Environmental Sciences* 35: 6–14. [CrossRef]

Romero-Hernández, Omar, and Sergio Romero. 2018. Maximizing the value of waste: From waste management to the circular economy. *Thunderbird International Business Review* 60: 757–64. [CrossRef]

Rousseeuw, Peter. 1987. Silhouettes: A graphical aid to the interpretation and validation of cluster analysis. *Journal of Computational and Applied Mathematics* 20: 53–65. [CrossRef]

Schneider, Petra, Le Hung Anh, Joerg Wagner, Jan Reichenbach, and Anja Hebner. 2017. Solid waste management in Ho Chi Minh City, Vietnam: Moving towards a circular economy? *Sustainability* 9: 286. [CrossRef]

Zeller, Vanessa, Edgar Towa, Marc Degrez, and Wouter Achten. 2019. Urban waste flows and their potential for a circular economy model at city-region level. *Waste Management* 83: 83–94. [CrossRef] [PubMed]

Zohoori, Mahmood, and Ali Ghani. 2017. Municipal Solid Waste Management Challenges and Problems for Cities in Low-Income and Developing Countries. *International Journal of Science and Engineering Applications* 2: 39–48. [CrossRef]

© 2019 by the authors. Licensee MDPI, Basel, Switzerland. This article is an open access article distributed under the terms and conditions of the Creative Commons Attribution (CC BY) license (http://creativecommons.org/licenses/by/4.0/).

Article

Tourism-Based Circular Economy in Salento (South Italy): A SWOT-ANP Analysis

Pasquale Marcello Falcone

Bioeconomy in Transition Research Group, IdEA, Unitelma Sapienza—University of Rome, Viale Regina Elena, 291, 00161 Roma, Italy; pasquale.falcone@unitelmasapienza.it

Received: 11 June 2019; Accepted: 10 July 2019; Published: 16 July 2019

Abstract: This paper is aimed at eliciting, by means of a multi-level perspective, potential drivers and barriers of the tourism industry in order to generate valuable information for policy makers to improve policy strategies for an effective transition towards sustainability. A Strengths, Weaknesses, Opportunities and Threats–Analytic Network Process (SWOT-ANP) framework was employed to explore the potential development of a second-generation biorefinery in Salento (a touristic area located in the southeast of Italy in Apulia Region) able to integrate waste management, renewable energy and bio-products production based on resource circularity in the tourism industry. Results indicate that survey participants recognized a higher level of priority for the pressures coming from the overall external setting involving values, dominant practices, rules and technologies (landscape and regime) over the internal tourism industry dynamics (niche). Results also show that the top five ranked factors are mainly pertaining to weaknesses (excessive bureaucracy and lack of technology and infrastructure) and threats (social acceptability and lack of long-term planning by governments), which can concretely jeopardize the transition towards a greater sustainability in the investigated area. The analysis presented constitutes a valuable model for agenda setting in order to find adequate policy actions to promote the transition.

Keywords: tourism industry; Italy; circular economy; multi-level perspective; SWOT

1. Introduction

Tourism is one of the key sectors for the socio-economic development of many countries worldwide (Muñoz and Navia 2015). Touristic regions are characterized by waste abundancy, given their high population density. Specifically, the average amount of municipal solid waste (thereafter MSW) produced by each of about 500 million citizens of the European Union was equal to 477 kg per year in 2015 (EUROSTAT 2017). Against this background, promoting a more sustainable economy, where production is obtained with fewer inputs, less waste and less greenhouse gas emissions represents a fundamental step towards appropriate waste management (D'Adamo 2018). From a policy perspective, in January 2018, the European Commission adopted an ambitious Circular Economy Package (European Commission 2018) which includes, among others, legislative schemes to improve the diffusion of innovative technologies, increase energy efficiency, reduce the dependence on imported raw materials and provide economic opportunities and long-term profitability (Morone et al. 2019). The circular economy (CE) approach has the goal of making better use of resources/materials through reuse, recycling and recovery in order to minimize the energy and environmental impact of resource extraction and processing (Ardolino and Arena 2019). This goal is mainly pursued by redesigning the life cycle of the product, with the aim of having minimal input and minimal production of system waste (D'Amato et al. 2017).

Tourism is one of the fastest growing industries in the world and one of the most remarkable socio-economic phenomena of the current era, and it represents an important determinant of waste generation (Arbulú et al. 2015). From a circular perspective, waste generated by the tourism industry can be, if properly managed, used as a resource for the city system and thus be a part of the urban processes able to optimize the rate of resource utilization (Girard and Nocca 2017). A way to approach part of the environmental pressure arising from tourism is the transition towards a holistic planning and design of integrated MSW processing activities by means of urban biorefineries, which are able to close the resource loop and increase resource efficiency (Satchatippavarn et al. 2016). Despite the agri-food sector still being the most predominant (Ronzon and M'Barek 2018), biorefining, i.e., the sustainable processing of biomass into a spectrum of marketable products and energy, is rapidly gaining ground and already represents an important part of the Italian bioeconomy (Intesa San Paolo 2018). In this framework, the biorefineries located at Gela (South Italy) and Porto Marghera (North Italy) represent a concrete example of a sustainable transition from fossil fuel to bio-based technologies, making possible the use of second-generation raw materials (e.g., palm oil, food waste, animal fats, etc.).

Biorefinery has emerged as a potential alternative to petroleum-based refinery, where biomass of non-edible waste is used as raw material and a range of products, such as biofuel, industrial biochemicals and biomaterials including commercially important biopolymers, are produced from a CE perspective (Clark and Deswarte 2015). However, an overall lack of social acceptability could be a significant barrier to the development of bioenergy-related industries (McGuire et al. 2017). For this transition to happen, it will require a joint effort by all concerned parties; it is not enough just to use biomass for industrial applications or to employ renewable resources and waste instead of fossil-based materials (Falcone 2018a). To meet this challenge, a transition must also take place from a socio-cultural point of view, stimulating local communities' awareness, enhancing dialogue among involved stakeholders, and benefitting from a proactive local policy making (Ehnert et al. 2018).

The environmental impacts of MSW production created pressures on public authorities to develop policy actions and strategies to deal with this concern (Lundmark and Stjernström 2009). The analysis of these strategies and their effect is particularly relevant for tourism destinations, since tourism inflows create an extra source of MSW and the attractiveness of a tourism destination can be disturbed by waste management (Arbulú et al. 2017). The limitation on land in certain tourism areas, the increasing costs of landfilling along with the necessity to safeguard the destination image have made waste management in touristic destinations particularly complex (Arbulú et al. 2016; Gómez et al. 2008).

The transformation of tourism toward sustainability requires a cross-disciplinary approach whose circular principles should include (Pan et al. 2018): (i) new models of production and consumption in order to minimize waste and convert wastes into valuable products, (ii) using biodegradable products for guests, (iii) creation of cultural values, such as conserving cultural heritage and traditional values, (iv) greening the tourism industry by creating conditions for enabling tourism operators to make long-term investments.

Building on these assumptions, this study tries to complement the recent interest towards tourism and circularity principles by investigating how local stakeholders would perceive and support fundamental changes in structures, cultures and practices for a sustainability transition towards an advanced biorefinery in Salento (a touristic area located in the southeast of Italy in Apulia Region) able to integrate renewable energy management and waste management in the industry. Sustainability transitions are long-standing, multi-dimensional, and essential transformation processes through which traditional socio-technical systems move towards new and more sustainable approaches of consumption and production (Loorbach and Rotmans 2010). The multi-level perspective (MLP) (Geels 2005), which considers the socio-technological system as characterized by three interacting layers (i.e., micro, meso and macro), is one of the main theoretical approaches to frame this change. However, there is a gap of research in understanding the actual dynamics with respect to the transition towards a tourism-based circular economy where smaller configurations (value-based and place-based) for sustainability might act as agents of change, enabling the transition process (Schäpke et al. 2017). With this in mind, the present study embraces a holistic approach, integrating the MLP with a Strengths, Weaknesses, Opportunities and Threats Analytic Network Process (SWOT-ANP) framework to exploit local stakeholders' knowledge and perspectives in order to generate valuable information for policy makers and propose an agenda setting of policy actions to promote the transition. Therefore, the research question is the following:

RQ: *Identifying the weaknesses, strengths, opportunities, and threats characterizing the tourism industry in Salento in order to propose an agenda setting of policy actions relevant for the transition towards circularity.*

The remainder of the paper is organized as follow: Section 2 sets out the theoretical framework; Section 3 introduces the case study and the methods employed; Section 4 presents and discusses the main findings; Section 5 provides some concluding remarks.

2. Theoretical Framework

The assumptions underlying the neo-Schumpeterian evolutionary theory of technological change reflect the need for a comprehensive development of two interdependent sub-systems: the techno-economic and the socio-institutional (Pérez 2010). In this vein, technological change represents just one aspect to consider together with social and institutional transformations. For a transition to occur, actors involved at technological, social and institutional levels must look at the same direction and share a common vision of the future (Fischer and Newig 2016).

In this fashion, we should refer, alongside the technological dimension of an innovation path, to societal transition that embrace both changes from bottom-up (e.g., user practices) and top-down perspectives (e.g., regulatory and institutional). The topic concerning how to promote and steer a sustainability transition has gathered growing interest among scholars, practitioners, and policymakers (Frantzeskaki and Loorbach 2010; Smith et al. 2005). A sustainability transition refers to a fundamental transformation in the system configuration towards a more sustainable option (Geels 2002). It is characterized by long-term, multi-dimensional, and fundamental transformation processes through which established socio-technical systems shift to more sustainable modes of production and consumption (Markard et al. 2012). Transitions towards sustainability succeed whenever there is a technological niche sufficiently developed coupled with adequate pressure arising from the landscape level (Lopolito et al. 2011). The MLP is one of the main conceptual approaches to frame this change. Essentially, in the MLP, transitions occur as a result of the interface between three different levels: landscape, regime and niches (Geels and Schot 2007). The landscape level embodies exogenous determinants including material and social infrastructure, politics, natural setting, etc. The regime represents a stable set of institutional rules, technical knowledge, and social interaction patterns shaping the fundamental configuration of technologies. Finally, the innovation niche can be conceived as a protected space where promising technologies are developed and experimented. Landscape factors could exert pressure on the incumbent regime and open windows of opportunities for niches to break through and conduce to radical shifts in socio-technical regimes (Geels 2011). A sustainable transition occurs from the interaction of the MLP levels, namely, when a sufficiently developed niche-innovation challenges the dominant regime which, in turn, undergoes an adequate amount of pressure from the landscape (Hansen and Nygaard 2014). However, opportunities for niche innovations to emerge and replace the current regime might be heavily hampered by external factors at the regime level and internal factors at the niche level (Falcone et al. 2019). Following the MLP, we build the analysis upon a combined SWOT-MLP framework to provide crucial theoretical perceptions for the transition under investigation. The main objective of SWOT analysis is to inspect internal and external system characteristics simultaneously, with the aim of supporting operational actions (Kurttila et al. 2000). SWOT analysis encompasses two main types of factors influencing the investigated system: internal factors (i.e., strengths and weaknesses) and external factors (i.e., opportunities and threats). Therefore, two conditions—one internal and one external—should be met for the transition towards circularity to succeed:

(i) the internal condition concerns the niche development (i.e., the second-generation biorefinery), whose strengths and weaknesses represent the extent to which the niche is effectively mature and thus ready for breakthrough;

(ii) the external condition can be assumed as a mix of regime and landscape pressures that act as opportunities and threats surrounding the tourism sector and support or hinder the transition. Within this operational framework, we can draw potential new strategies for an effective transition towards a tourism-based circular economy (see Figure 1).

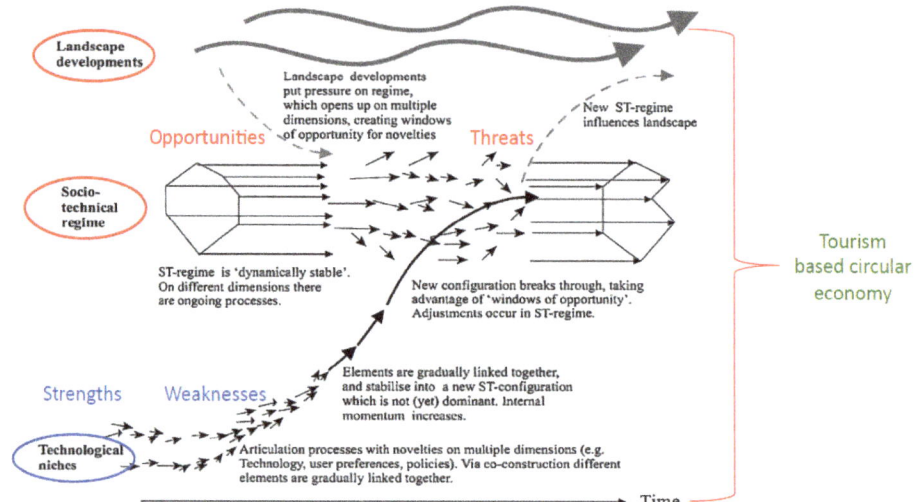

Figure 1. Strengths, Weaknesses, Opportunities and Threats - Multi-Level Perspective (SWOT-MLP) for a transition towards a tourism-based circular economy. Source: adapted from Falcone et al. (2019) and Geels (2011).

The MLP is a valuable tool for policy makers to understand and thus to address transitions in an efficient and effective way by placing the focus on both niche and regime levels (Coenen et al. 2010). Therefore, it allows us to gain an in-depth understanding of the framework conditions surrounding the tourism industry in Salento. With this understanding, potential policy directions for an effective transition towards a tourism-based circular economy can be suggested.

3. Materials and Methods

3.1. The Study Area

The Salento peninsula is an area located in the southeast of Italy in Apulia Region (Figure 2). It covers the provinces of Lecce, Taranto (the eastern part) and Brindisi (central-southern part), with a population of approximately 1.5 million inhabitants. Salento spans 5329 km^2 and has more than 300 km of coastline along the Ionic and Adriatic Seas.

In recent decades, Salento became known at both national and international levels due to its beautiful coasts and numerous events and entertainments proposed along with its well-known historical and artistic heritage. The economy, once purely based on agriculture and artisanal fisheries, has experienced a significant increase in the secondary and tertiary sectors, making this area one of the richest in Southern Italy. One of the most important economic sectors is tourism. Tourist arrivals registered an 80% increase between 2002 and 2009, followed by a more moderate 10% increase between 2009 and 2015 (ISTAT 2016). The tourism industry is especially intensive in MSW generation compared to other economic sectors, such as manufacturing or agriculture, and more prone to produce other kinds of polluting outputs (Mateu-Sbert et al. 2013). Therefore, the relationship between tourism growth and MSW generation in Salento and related management strategies are worth studying for at least three reasons: (i) the development of the tourism industry has resulted in an increase in waste generation

(UNEP/GPA 2006); (ii) inadequate MSW management can bring negative effects on the attractiveness of the touristic area, reducing tourism inflows (Arbulú et al. 2015); (iii) valorizing available waste materials, in circular economy models, allows for closing the loop not only material-wise but also energy-wise (Pan et al. 2018). Due to these premises, the tourism industry in Salento represents an interesting case of investigation for understanding the socio-political dynamics based on experts' insights and awareness in order to support a radical form of sustainability transition. Italy has recently taken steps in this direction with two second-generation biorefineries, namely Gela and Porto Marghera. However, they still represent a small industrial niche, facing strong socio-economic challenges (Imbert et al. 2017). From this perspective, the tourism industry in Salento could represent an open-air laboratory for the application of the most advanced environmental and renewable technologies and become a frontrunner not only for Italy but also for the whole EU.

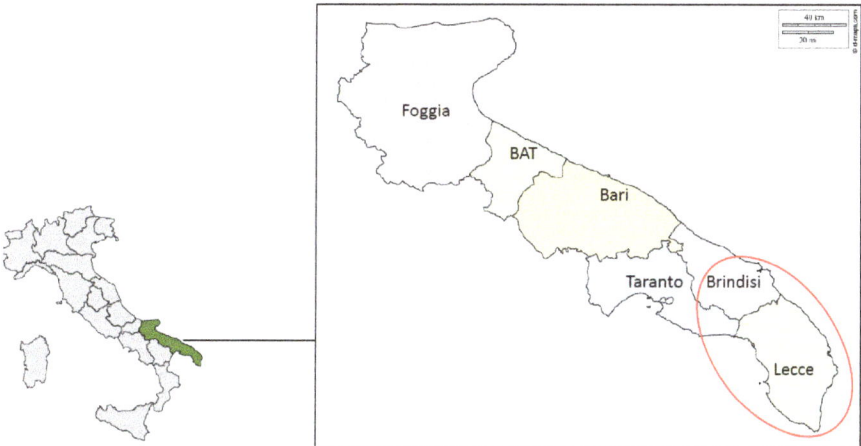

Figure 2. Geographical position of the study area. **Source:** author.

3.2. Methods

With the aim of determining the quantitative values of SWOT factors, the Analytic Hierarchy Process (AHP) or the ANP are the most suitable techniques (Saaty 1996). The AHP is generally employed to determine the quantitative values for SWOT analysis, since it works on the idea that elements function independently of one another in a hierarchical configuration (Catron et al. 2013; Saaty 2005). This represents a severe assumption to meet, especially when the considered attributes become interdependent owing to a complex situation. Such a degree of complexity makes the ANP appropriate to study factors' dependencies (Starr et al. 2019). Assessing the conditions able to promote socio-institutional changes for a sustainable energy transition in Salento (e.g., second-generation biorefinery) includes a number of complexities and interdependencies involving different stakeholders. For example, urbanization, demographic trends and related socio-cultural changes will likely impact the waste management practices in the area. Accordingly, SWOT-ANP is the appropriate method for this analysis.

The methodological approach can be divided in two distinct phases:

1. Identification and selection of relevant SWOT factors by means of a literature review and expert interviews;
2. Prioritization of the internal and external factors identified through a survey administered to a variety of knowledgeable stakeholders.

In the first phase, a literature review was carried out by looking at two main databases of scientific literature, i.e., Scopus and Web of Science, to ascertain a list of relevant factors to be used in our investigation. A broad keyword search was conducted in order to retrieve relevant papers within the publication timeframe of 2015–2019. We paired some anchor keywords (i.e., "bio*," "circular*," and "sustainab*") with search strings (i.e., "tourism", "energy", "transition", "refinery"). Our in-depth literature review uncovered more than 150 papers engaging with the sustainability of the tourism sector and more than 20 regarding waste-to-energy transitions. With the aim of selecting the most relevant factors, we refined this pool of articles by carefully examining the text of each article in order to ascertain the presence of a well-defined idea or value judgement with regard to the area of investigation. In doing so, we employed the QDA Miner 5.0 software package (Provalis Research 2015), which allowed us to perform a qualitative assessment of the context in which relevant keywords appeared in the selected documents. Table 1 reports some descriptive statistics about the documents analyzed and the relative keywords found.

Table 1. Descriptive statistics of the keywords search.

Total number of documents analyzed	151
Total number of words	1,583,418
Average number of words per document	10,486
Total number of keywords found	847
Average number of keywords per document	5.6

We accessed 151 documents for a total of 1,583,418 words, with an average number of words per document equal to 10,486. The majority of documents refer to the sustainability and development of the tourism sector. In this framework, we found 847 keywords corresponding to an average of 5.6 per each accessed document. The word map below (Figure 3) allows the visualization of the most relevant words employed in literature to characterize the sustainability of the tourism industry.

From the total number of words used in the 151 articles, the map includes only those terms which appear at least 5 times in the analyzed corpus of the single article. The bigger the letter size, the more frequent the word. It is important to mention that the term energy is in the middle of the map, and it is connected to all the focal points of the current research (biorefinery, development, tourism, circular), but also to other important aspects, such as: transition, policy, jobs, etc. In a further stage, with the aim of choosing the most relevant factors and summarizing them by way of a 2 × 2 matrix (internal factors: strengths and weaknesses; vs external factors: opportunities and threats), we conducted two interviews with two academicians (i.e., an agricultural economist and a commodity scientist) with long-term involvement (i.e., more than a decade) in the field under investigation. This allowed the labelling of the factors retrieved by means of literature review as internal and external to the tourism industry.

In the second phase, and building on the protocol followed in Starr et al. (2019), a survey was developed and administered to a group of knowledgeable local stakeholders. A larger group of experts was identified, starting from a preliminary list of actors derived from the Italian Association of Tourism Professionals and Cultural Operators (AIPTOC). The association has more than 300 effective members covering different categories (e.g., managers, researchers, evaluators of management systems, consultancy companies, institutions and trade associations, etc.). Successively, considering the information collected by means of websites, technical reports and blogs, we refined the list by focusing only on actors with long-term involvement (i.e., more than a decade) in the field under investigation for the selected study area. Interviewees were selected with the intention of representing a wide range of actors involved in the tourism industry. In particular, the group of experts taking part in the survey were: two tourism industry professionals, a trade association, two representatives of a consumer association and environmental associations, a local policy maker, and two researchers. The eight interviews were conducted by telephone over the period of February to April 2019, and lasted approximately one hour. Respondents were asked to make several pairwise comparisons between the identified SWOT factors using a scale suggested by Saaty (1996). The scale ranges from equal importance (participant assigns a numerical value of 1) to extreme importance (participant assigns a numerical value of 9) of one element over another. After comparisons between each factor within the SWOT categories were made, comparisons between each category were made by employing the same protocol. Therefore, two matrices were administered to respondents: (i) pairwise comparisons per group (Table 2); and (ii) pairwise comparisons of the groups (i.e., strengths, weaknesses, opportunities and threats) (Table 3).

Figure 3. Word map of text-based analysis of the tourism industry. **Source:** author.

Table 2. Pairwise comparisons per group.

	F1	F2	F3	F4	F5
F1	1	$1/V_{F2F1}$	$1/V_{F3F1}$	$1/V_{F4F1}$	$1/V_{F5F1}$
F2	V_{F2F1}	1	$1/V_{F2F3}$	$1/V_{F2F4}$	$1/V_{F2F5}$
F3	V_{F3F1}	V_{F2F3}	1	$1/V_{F3F4}$	$1/V_{F3F5}$
F4	V_{F4F1}	V_{F2F4}	V_{F3F4}	1	$1/V_{F4F5}$
F5	V_{F5F1}	V_{F2F5}	V_{F3F5}	V_{F4F5}	1
Sum	SC_{F1}	SC_{F2}	SC_{F3}	SC_{F4}	SC_{F4}

Table 3. Pairwise comparisons of the groups.

	G1	G2	G3	G4
G1	1	$1/V_{G2G1}$	$1/V_{G3G1}$	$1/V_{G4G1}$
G2	V_{G2G1}	1	$1/V_{G2G3}$	$1/V_{G2G4}$
G3	V_{G3G1}	V_{G2G3}	1	$1/V_{G3G4}$
G4	V_{G4G1}	V_{G2G4}	V_{G3G4}	1
Sum	SC_{G1}	SC_{G2}	SC_{G3}	SC_{G4}

In Tables 2 and 3, F1,..., F5 are the identified factors, G1, ..., G4 are the SWOT groups, V_{F2F1} represents the value of factor F2 with respect to factor F1 (the same logic is applied to all factors), V_{G2G1} is the value of group G2 with respect to group G1 (the same logic is applied to all groups), SC_{F1} and SC_{G1} are the sum of values regarding the columns of group F1 and G1, respectively (the same logic is applied to all groups)[1].

The local factor priority is obtained evaluating the average values of the expert comparisons among the factors in the same SWOT group. Meanwhile, the group priority is based on the average of the expert comparison among all groups. The global factor priority of all SWOT factors is calculated as the product of the local factor priority and the respective group priority.

The two aforementioned phases are common in both AHP and ANP procedures. However, to appraisal the interdependence between SWOT factors, an additional analysis is needed. Table 4 was employed to weight the interdependence of each category. For example, respondents were asked to consider how strengths may be used to mitigate weaknesses or enhance opportunities (Catron et al. 2013; Starr et al. 2019).

Table 4. Assessing the interdependence of each category.

	G1	G2	G3	G4
G1	1	I_{G1G2}	I_{G1G3}	I_{G1G4}
G2	I_{G2G1}	1	I_{G2G3}	I_{G2G4}
G3	I_{G3G1}	I_{G3G2}	1	I_{G3G4}
G4	I_{G4G1}	I_{G4G2}	I_{G4G3}	1

In Table 4, I_{G2G1}, I_{G3G1} and I_{G4G1} represent the factors' interdependence. They measure the relative importance of weaknesses, opportunities, and threats in enhancing strengths, respectively, and I_{G1G2}, I_{G3G2} and I_{G4G2} represent the relative importance of strengths, opportunities, and threats relative to mitigating weaknesses and so forth.

The global priority value based on factor interdependence, for individual SWOT factors, can then be calculated as: global priority of factor G_{ij} = priority value of factor G_{ij} * (interdependent scaling value of SWOT category).

4. Results and Discussion

The first step of the methodological approach allows the identification and selection of the relevant SWOT factors describing the tourism industry, with particular emphasis on the investigated area. According to the literature review and experts' perspectives, several driving forces and barriers might trigger or hamper the transition towards a tourism-based circular economy in Salento (Table 5).

The mere analysis of different SWOT factors highlighted that a possible driving force (i.e., opportunities and/or threats) of the tourism industry relates to the potential engagement of several knowledgeable stakeholders in sustainable production and consumption processes. The scientific and technological collaboration among actors along the whole tourism chain could ensure both environmental sustainability and social inclusion (Kohon 2018). On the other hand, possible barriers (i.e., weaknesses and threats) to the development of a second-generation biorefinery encompass institutional (e.g., policy uncertainty), financial (e.g., low financial support) and social factors (low social acceptance). These findings are in line with the literature, suggesting that one of the main concerns of local government officials is the potential negative effect of new projects and plants on the tourism industry (Vargas Payera 2018). This could explain the lack of long-term planning by policy makers, whose attention are sometimes mainly directed towards the short-term consensus (Laird 2001).

---

[1] Following Margles et al. (2010), the obtained results have been tested for consistency.

Table 5. Strengths, Weaknesses, Opportunities and Threats (SWOT) analysis of the tourism industry in the study area.

Strengths		Weaknesses	
S1	High number of involved actors	W1	Limited sectorial expertise
S2	Utilization of non-marketable waste	W2	Low financial support
S3	Technical requirements well-known	W3	Lack of awareness
S4	Production of value-added products	W4	Lack of technology and infrastructure
S5	Additional source of income	W5	Excessive bureaucracy
Opportunities		Threats	
O1	Pollution reduction and land remediation	T1	Lack of long-term planning
O2	Reduced dependency on energy imports	T2	Social acceptability (Not In My Back Yard, NIMBY)
O3	Building infrastructure	T3	Policy uncertainty
O4	Scientific and technological collaboration	T4	Poor attitude towards waste management
O5	Increasing green jobs	T5	Competition from other energy sources

Following the SWOT-MLP combined framework (Falcone et al. 2019), two conditions—one internal (i.e., niche development) and one external (i.e., regime and landscape pressures)—should be met for the transition towards circularity to happen. Analyzing the stakeholders' perspective towards the prioritization of the SWOT factor provides some preliminary information on the effective possibility for the niche to breakthrough (Figure 4).

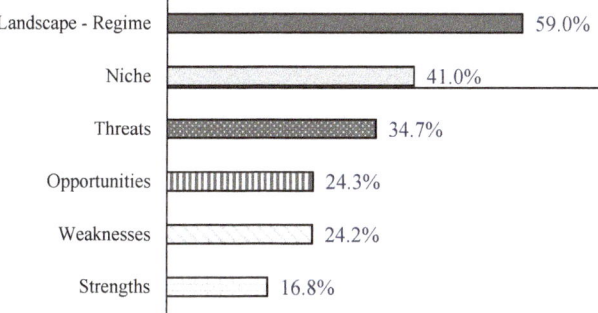

Figure 4. Priority levels of SWOT-MLP factors. **Source:** author.

Specifically, the aggregation of different levels of priority assigned by respondents to each SWOT category allows us to assign the highest priority to threats (34.7%), followed by opportunities (24.3%) and weaknesses (24.2%), while strengths are recognized with a lower priority level (16.3%). Taking together internal niche factors (i.e., strengths and weaknesses) and external regime and landscape pressures (i.e., threats and opportunities), experts recognized a higher level of priority for the external pressures (59%) over the internal niche dynamics (40%).

As mentioned in the previous section, the global factor priority of all SWOT factors is calculated as the product of the local factor priority and the respective interdependent group priority. The values of global priorities of the SWOT factors, determined through ANP, as well as their priority rankings are shown in Table 6. Figure 5 provides a graphical representation. Factors further away from the origin are relatively more important than factors closer to the origin.

Table 6. The global priorities for SWOT factors determined through Strengths, Weaknesses, Opportunities and Threats–Analytic Network Process (SWOT-ANP).

Strengths		Global Priority	Global Ranking
S1	High number of involved actors	0.022	18
S2	Utilization of non-marketable waste	0.028	15
S3	Technical requirements well-known	0.017	19
S4	Production of value-added products	0.031	13
S5	Additional source of income	0.041	8
Opportunities			
O1	Pollution reduction and land remediation	0.046	6
O2	Reduced dependency on energy imports	0.037	11
O3	Building infrastructure	0.039	9
O4	Scientific and technological collaboration	0.011	20
O5	Increasing green jobs	0.058	3
Weaknesses			
W1	Limited sectorial expertise	0.045	7
W2	Low financial support	0.033	12
W3	Lack of awareness	0.026	16
W4	Lack of technology and infrastructure	0.047	5
W5	Excessive bureaucracy	0.059	2
Threats			
T1	Lack of long-term planning	0.049	4
T2	Social acceptability (NIMBY)	0.065	1
T3	Policy uncertainty	0.037	10
T4	Poor attitude towards waste management	0.029	14
T5	Competition from other energy sources	0.025	17

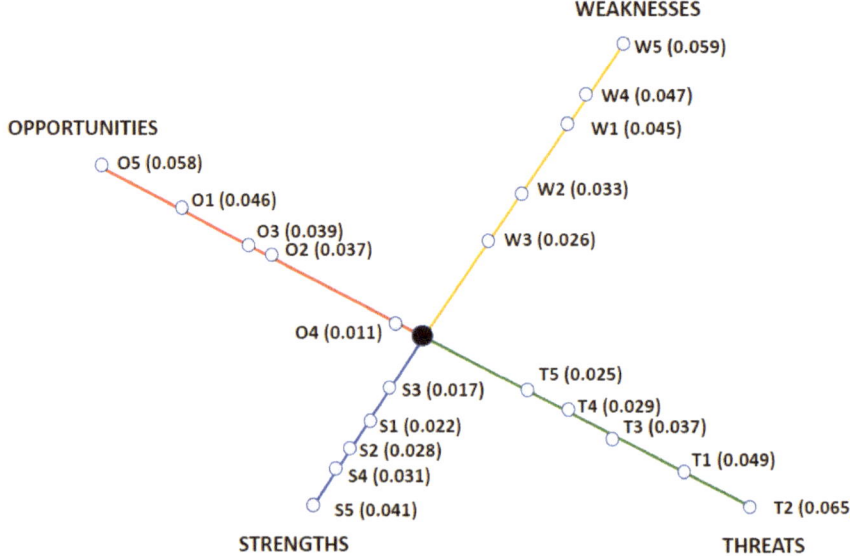

Figure 5. Graphical representation of SWOT factors and their corresponding global priorities. Source: author.

For strengths, respondents recognized that with the installation of a biomass pre-treatment plant, and using second-generation raw materials derived from non-marketable waste (S2), a second-generation biorefinery could be also able to produce an additional source of income (S5), and this represents a circular economy model. A clear example is represented by the recovery of organic waste from accommodation facilities towards responsible initiatives. Literature has documented the relevance of responsible tourism in the tourism industry (Ruiz-Lozano et al. 2018; Wocke and Merwe 2007). It emphasizes practices that are environmentally friendly, socially acceptable and economically beneficial to all stakeholders (Musavengane and Steyn 2013). On the supply side, hotels should focus on green purchasing, ecolabelling and certification, waste management and recycling (Mensah and Blankson 2014). Waste disposal should be phased out and, where it is unavoidable, policy makers need to develop sustainable strategies appropriate for the community need to be accomplished with responsible practices (Goffi et al. 2019). Industry operators can be encouraged by the public sector to participate in responsible tourism, including education, economic motivation, marketing motivation and building social networks (Musavengane 2019), in order to sufficiently increase public awareness and to correct the public perception of unsustainable tourism practice towards an environmental practice (Ruban et al. 2019). From this perspective, the existence of devices in the tourist's home is a determining factor for acquiring digital knowledge, skills and attitudes (Díaz-Meneses 2019).

With reference to weaknesses, the analysis emphasized that the administrative burdens of bureaucracy (W5) was the top-rated concern. Excessive bureaucracy in Italy has caused a competitive disadvantage compared to other EU countries in attracting private investments, including in the tourism industry (Falcone 2018b). Bureaucracy represents a classic issue in change management (Bevir 2009; Hall 2005). However, there is still little empirical evidence regarding the issue of bureaucracy in the tourism industry. In a recent study on the effect of bureaucracy on the tourism sector in Italy, Marino and Pariso (2018) contribute to creating a pool of knowledge related to change management by underlining the relevant elements within different bureaucratic typologies. As found by the authors, the private sector is better than the public sector in some specific ways: private sector organizations are more cost conscious, more inclined to implement modern personnel management and more capable of developing corporate change as a steering instrument. Therefore, pointing at the involvement of private actors could have more effective results in the transition towards sustainability. Moreover, the lack of technology and infrastructure (W4) for a proper waste management has been recognized as an important weakness among the majority of stakeholders. They also expressed concern about the low financial support (W2) mainly due to: (i) limited sectorial expertise of the potential investors; and (ii) short term orientation of financial tools. Financial instruments must be matched with the development of science and technology progress and the financing needs of renewable energy; energy security law and energy funds accord with political objectives in order to better promote the development of the waste-to-energy industry (Wang and Zhi 2016). However, touristic businesses operating in rural areas face, overall, higher difficulties in accessing finances compared to similar businesses located in more industrialized areas (Badulescu et al. 2015).

With respect to opportunities, interviewees recognize a proactive role for local policy makers in incentivizing biorefinery development. The construction of infrastructure (O3) was identified as an important opportunity for biorefinery development in Salento by providing job opportunities for local population (O5). This action includes donor funds aimed at the installation of new environmentally friendly plants (i.e., biorefinery facilities, R&D center), infrastructural subsidies (e.g., storage platforms for biomass serving the biorefinery) and long-term assets (i.e., transportation, energy and social infrastructures). As recognized by respondents, large infrastructural investments can foster local economic development along the whole supply chain by increasing the firms' economic performance. This finding is supported by the literature (Bostick et al. 2018; Falcone et al. 2017). Moreover, increasing efficiency in the industry is expected to have positive effects on local employment; as such, infrastructure investments are likely to impact on quality of life and well-being.

For threats, the respondents highly recognized the relevance of the social acceptability of a new biorefinery plant (T2). Specifically, they pointed to the lack of a well diffused environmental culture in the local community as a possible obstacle. For example, with reference to waste management and NIMBY (Not In My Back Yard) attitudes, citizens do not want plants situated within their cities, because of the potential issues for the pollutants and odor which might have negative effects on the tourism industry (D'Adamo et al. 2019). It is important to ensure social sustainability by providing a healthy and safe environment for all stakeholders, in both physical and psychological aspects (Zuo and Zhao 2014), not only for the current time but also for future development (Lu et al. 2019). Similarly, a lack of long-term planning by governments (T1) and overall policy uncertainty (T2) also received a relatively high ranking from all the stakeholders within the region. Specifically, policy uncertainty distorts the fundamental relation between investment and the cost of capital (Drobetz et al. 2018) and is a significant challenge for actors in the renewable energy sector (Dalby et al. 2018). As a starting point, including biorefineries in the government's long-term strategic plans is a relevant way to pave. It could guide and reassure investors, providing, thus, capital to deploy commercial-scale versions of mature biorefinery technologies (Ellen MacArthur Foundation 2015).

5. Conclusions

Transitioning from a fossil fuel-based economy to one based on the use of biomass is increasingly perceived as a needed feat among scholars, analysts and policy makers. Defining effective ways to align sustainable supply chain practices to the CE paradigm represents a cutting edge topic at the intersection of scientific research and public policy (Genovese et al. 2017).

In this framework, the present paper has shed light on external pressures and internal dynamics so as to provide a clear direction for policy strategies to support the transition towards a tourism-based circular economy. To this aim, we built our analysis upon an integrated SWOT-MLP framework to provide crucial theoretical perceptions for the transition under investigation.

The findings emerging from our investigation can represent useful insights for policy makers. Specifically, among the top five ranked factors, we can recall: (i) T2 (social acceptability); (ii) W5 (excessive bureaucracy); (iii) O5 (green jobs); (iv) T1 (lack of long-term planning by governments); (v) W4 (lack of technology and infrastructure). Four out of these five factors represent weaknesses (i.e., W4, W5) and threats (i.e., T1, T2), which can concretely jeopardize the transition towards greater sustainability in the investigated area.

Policy strategies should aim at reducing the administrative burdens of bureaucracy by introducing, for example, on a large scale, e-government services. Moreover, promoting information campaigns could increase the degree of social awareness and reduce the NIMBY concern. Additionally, supporting public infrastructural investments can raise local economic development and increase firms' economic performance so as to exploit the potential creation of green jobs.

The main limitation of this approach rests on the qualitative nature of the methodological approach, which is not able to identify the most effective policy strategies (policy design) nor to appraise the financial support for each measure (policy engineering). Nevertheless, this approach is crucial to provide a clear direction for policy maker interventions. Further lines of research could aim at extending this investigation to policy design, by including a fuzzy inference simulation based on a causal-effect map, to identify the most effective instrument mix for the development of the tourism-based circular economy.

Funding: This research received no external funding.

Acknowledgments: The author gratefully acknowledge the great support of four anonymous reviewers and the guest editor for their careful reading of the manuscript and their many insightful comments and suggestions.

Conflicts of Interest: The author declares no conflict of interest.

References

Arbulú, Italo, Javier Lozano, and Javier Rey-Maquieira. 2015. Tourism and Solid Waste Generation in Europe: A Panel Data Assessment of the Environmental Kuznets Curve. *Waste Management* 46: 628–36. [CrossRef] [PubMed]

Arbulú, Italo, Javier Lozano, and Javier Rey-Maquieira. 2016. The Challenges of Municipal Solid Waste Management Systems Provided by Public-Private Partnerships in Mature Tourist Destinations: The Case of Mallorca. *Waste Management* 51: 252–58. [CrossRef] [PubMed]

Arbulú, Italo, Javier Lozano, and Javier Rey-Maquieira. 2017. The Challenges of Tourism to Waste-to-Energy Public-Private Partnerships. *Renewable and Sustainable Energy Reviews* 72: 916–21. [CrossRef]

Ardolino, Filomena, and Umberto Arena. 2019. Biowaste-to-Biomethane: An LCA Study on Biogas and Syngas Roads. *Waste Management* 87: 441–53. [CrossRef] [PubMed]

Badulescu, D., Adriana Giurgiu, Nicolae Istudor, and Alina Badulescu. 2015. Rural Tourism Development and Financing in Romania: A Supply-Side Analysis. *Agricultural Economics (Zemědělská Ekonomika)* 61: 72–82. [CrossRef]

Bevir, Mark. 2009. *Key Concepts in Governance*. Chennai: Sage. [CrossRef]

Bostick, Thomas, Connelly Elizabeth, Lambert James, and Linkov Igor. 2018. Resilience Science, Policy and Investment for Civil Infrastructure. *Reliability Engineering & System Safety* 175: 19–23. [CrossRef]

Catron, Jonathan, G. Andrew Stainback, Puneet Dwivedi, and John M. Lhotka. 2013. Bioenergy Development in Kentucky: A SWOT-ANP Analysis. *Forest Policy and Economics* 28: 38–43. [CrossRef]

Clark, James H., and Fabien Deswarte. 2015. *Introduction to Chemicals from Biomass*. Hoboken: John Wiley & Sons.

Coenen, Lars, Rob Raven, and Geert Verbong. 2010. Local Niche Experimentation in Energy Transitions: A Theoretical and Empirical Exploration of Proximity Advantages and Disadvantages. *Technology in Society* 32: 295–302. [CrossRef]

D'Adamo, Idiano. 2018. The Profitability of Residential Photovoltaic Systems. A New Scheme of Subsidies Based on the Price of CO2 in a Developed PV Market. *Social Sciences* 7: 148.

D'Adamo, Idiano, Pasquale Marcello Falcone, and Francesco Ferella. 2019. A Socio-Economic Analysis of Biomethane in the Transport Sector: The Case of Italy. *Waste Management* 95: 102–15. [CrossRef]

D'Amato, Dalia, Droste Niels, Allen Ben, Kettunen Marianne, Lähtinen Katja, Korhonen Jaana, Leskinen Pekka, Matthies Brent, and Toppinen Anne. 2017. Green, Circular, Bio Economy: A Comparative Analysis of Sustainability Avenues. *Journal of Cleaner Production* 168: 716–34. [CrossRef]

Dalby, Peder A. O., Gisle R Gillerhaugen, Verena Hagspiel, Tord Leth-Olsen, and Jacco J. J. Thijssen. 2018. Green Investment under Policy Uncertainty and Bayesian Learning. *Energy* 161: 1262–81. [CrossRef]

Díaz-Meneses, Gonzalo. 2019. A Multiphase Trip, Diversified Digital and Varied Background Approach to Analysing and Segmenting Holidaymakers and Their Use of Social Media. *Journal of Destination Marketing & Management* 11: 166–82. [CrossRef]

Drobetz, Wolfgang, Sadok El Ghoul, Omrane Guedhami, and Malte Janzen. 2018. Policy Uncertainty, Investment, and the Cost of Capital. *Journal of Financial Stability* 39: 28–45. [CrossRef]

Ehnert, Franziska, Niki Frantzeskaki, Jake Barnes, Sara Borgström, Leen Gorissen, Florian Kern, Logan Strenchock, and Markus Egermann. 2018. The Acceleration of Urban Sustainability Transitions: A Comparison of Brighton, Budapest, Dresden, Genk, and Stockholm. *Sustainability* 10: 612. [CrossRef]

Ellen MacArthur Foundation. 2015. *Delivering the Circular Economy: A Toolkit for Policymakers*. Ellen MacArthur Foundation.

European Commission. 2018. *Implementation of the Circular Economy Action Plan—Final Circular Economy Package*. Brussels: European Commission, Environment Website.

EUROSTAT. 2017. *477 Kg of Municipal Waste Generated per Person in the EU—Product*. Luxemburg: EUROSTAT.

Falcone, Pasquale Marcello. 2018a. Analysing Stakeholders' Perspectives towards a Socio-Technical Change: The Energy Transition Journey in Gela Municipality. *AIMS Energy* 6: 645–57. [CrossRef]

Falcone, Pasquale Marcello. 2018b. Green Investment Strategies and Bank-Firm Relationship: A Firm-Level Analysis. *Economics Bulletin* 38: 2225–39.

Falcone, Pasquale Marcello, Antonio Lopolito, and Edgardo Sica. 2017. Policy Mixes towards Sustainability Transition in the Italian Biofuel Sector: Dealing with Alternative Crisis Scenarios. *Energy Research & Social Science* 33: 105–14. [CrossRef]

Falcone, Pasquale Marcello, Almona Tani, Valentina Elena Tartiu, and Cesare Imbriani. 2019. Towards a Sustainable Forest-Based Bioeconomy in Italy: Findings from a SWOT Analysis. *Forest Policy and Economics*, 101910. [CrossRef]

Fischer, Lisa Britt, and Jens Newig. 2016. Importance of Actors and Agency in Sustainability Transitions: A Systematic Exploration of the Literature. *Sustainability (Switzerland)* 8: 476. [CrossRef]

Frantzeskaki, Niki, and Derk Loorbach. 2010. Towards Governing Infrasystem Transitions. *Technological Forecasting and Social Change* 77: 1292–301. [CrossRef]

Geels, Frank W. 2002. Technological Transitions as Evolutionary Reconfiguration Processes: A Multi-Level Perspective and a Case-Study. *Research Policy* 31: 1257–74. [CrossRef]

Geels, Frank W. 2005. The Dynamics of Transitions in Socio-Technical Systems: A Multi-Level Analysis of the Transition Pathway from Horse-Drawn Carriages to Automobiles (1860–1930). *Technology Analysis & Strategic Management* 17: 445–76.

Geels, Frank W. 2011. The Multi-Level Perspective on Sustainability Transitions: Responses to Seven Criticisms. *Environmental Innovation and Societal Transitions* 1: 24–40. [CrossRef]

Geels, Frank W, and Johan Schot. 2007. Typology of Sociotechnical Transition Pathways. *Research Policy* 36: 399–417. [CrossRef]

Genovese, Andrea, Adolf A. Acquaye, Alejandro Figueroa, and S. C. Lenny Koh. 2017. Sustainable Supply Chain Management and the Transition towards a Circular Economy: Evidence and Some Applications. *Omega (United Kingdom)* 66: 344–57. [CrossRef]

Girard, Luigi Fusco, and Francesca Nocca. 2017. From Linear to Circular Tourism. *Aestimum* 70: 51–74.

Goffi, Gianluca, Marco Cucculelli, and Lorenzo Masiero. 2019. Fostering Tourism Destination Competitiveness in Developing Countries: The Role of Sustainability. *Journal of Cleaner Production* 209: 101–15. [CrossRef]

Gómez, Carlos Mario, Javier Lozano, and Javier Rey-Maquieira. 2008. Environmental Policy and Long-Term Welfare in a Tourism Economy. *Spanish Economic Review* 10: 41–62. [CrossRef]

Hall, Colin Michael. 2005. The Role of Government in the Management of Tourism: The Public Sector and Tourism Policies. *The Management of Tourism*, 217–31.

Hansen, Ulrich Elmer, and Ivan Nygaard. 2014. Sustainable Energy Transitions in Emerging Economies: The Formation of a Palm Oil Biomass Waste-to-Energy Niche in Malaysia 1990–2011. *Energy Policy* 66: 666–76. [CrossRef]

Imbert, Enrica, Luana Ladu, Piergiuseppe Morone, and Rainer Quitzow. 2017. Comparing Policy Strategies for a Transition to a Bioeconomy in Europe: The Case of Italy and Germany. *Energy Research and Social Science* 33: 70–81. [CrossRef]

Intesa San Paolo. 2018. La Bioeconomia in Europa 4° Rapporto. Available online: http://www.instm.it/public/02/04/4°%20Rapporto%20sulla%20Bioeconomia.pdf (accessed on 22 April 2019).

ISTAT. 2016. Indicatori Demografici - Stime per l'anno 2015. Available online: www.istat.it (accessed on 22 April 2019).

Kohon, Jacklyn. 2018. Social Inclusion in the Sustainable Neighborhood? Idealism of Urban Social Sustainability Theory Complicated by Realities of Community Planning Practice. *City, Culture and Society* 15: 14–22. [CrossRef]

Kurttila, Mikko, Mauno Pesonen, Jyrki Kangas, and Miika Kajanus. 2000. Utilizing the Analytic Hierarchy Process (AHP) in SWOT Analysis—a Hybrid Method and Its Application to a Forest-Certification Case. *Forest Policy and Economics* 1: 41–52. [CrossRef]

Laird, Frank N. 2001. *Solar Energy, Technology Policy, and Institutional Values*. Cambridge: Cambridge University Press.

Loorbach, Derk, and Jan Rotmans. 2010. The Practice of Transition Management: Examples and Lessons from Four Distinct Cases. *Futures* 42: 237–46. [CrossRef]

Lopolito, Antonio, Gianluca Nardone, Maurizio Prosperi, Roberta Sisto, and Antonio Stasi. 2011. Modeling the Bio-Refinery Industry in Rural Areas: A Participatory Approach for Policy Options Comparison. *Ecological Economics* 72: 18–27. [CrossRef]

Lu, Jia-Wei, Yingshi Xie, Beibei Xu, Yuanqing Huang, Jing Hai, and Jixian Zhang. 2019. From NIMBY to BIMBY: An Evaluation of Aesthetic Appearance and Social Sustainability of MSW Incineration Plants in China. *Waste Management* 95: 325–33. [CrossRef]

Lundmark, Linda, and Olof Stjernström. 2009. Environmental Protection: An Instrument for Regional Development? National Ambitions versus Local Realities in the Case of Tourism. *Scandinavian Journal of Hospitality and Tourism* 9: 387–405. [CrossRef]

Margles, Shawn W., Michel Masozera, Louis Rugyerinyange, and Beth A. Kaplin. 2010. Participatory Planning: Using SWOT-AHP Analysis in Buffer Zone Management Planning. *Journal of Sustainable Forestry*. [CrossRef]

Marino, Alfonso, and Paolo Pariso. 2018. Italian Public Tourism Sector, Bureaucracy and Change Management Process: Four Bureaucratic Organizational Typologies. *Journal on Tourism & Sustainability* 2: 6–22.

Markard, Jochen, Rob Raven, and Bernhard Truffer. 2012. Sustainability Transitions: An Emerging Field of Research and Its Prospects. *Research Policy* 41: 955–67. [CrossRef]

Mateu-Sbert, Josep, Ignacio Ricci-Cabello, Ester Villalonga-Olives, and Elena Cabeza-Irigoyen. 2013. The Impact of Tourism on Municipal Solid Waste Generation: The Case of Menorca Island (Spain). *Waste Management* 33: 2589–93. [CrossRef] [PubMed]

McGuire, Julia B., Jessica E Leahy, James A. Marciano, Robert J. Lilieholm, and Mario F. Teisl. 2017. Social Acceptability of Establishing Forest-Based Biorefineries in Maine, United States. *Biomass and Bioenergy* 105: 155–63. [CrossRef]

Mensah, Ishmael, and Emmanuel J. Blankson. 2014. Commitment to Environmental Management in Hotels in Accra. *International Journal of Hospitality & Tourism Administration* 15: 150–71.

Morone, Piergiuseppe, Pasquale Marcello Falcone, and Antonio Lopolito. 2019. How to Promote a New and Sustainable Food Consumption Model: A Fuzzy Cognitive Map Study. *Journal of Cleaner Production* 208: 563–74. [CrossRef]

Muñoz, Edmundo, and Rodrigo Navia. 2015. Waste Management in Touristic Regions. *Waste Management & Research* 33: 593–94. [CrossRef]

Musavengane, Regis. 2019. Small Hotels and Responsible Tourism Practice: Hoteliers' Perspectives. *Journal of Cleaner Production* 220: 786–99. [CrossRef]

Musavengane, Regis, and Jacobus Nicolaas Steyn. 2013. Responsible Tourism Practices in the Cape Town Hotel Sub-Sector. *International Journal of Hospitality & Tourism Systems* 6: 52–63.

Pan, Shu-Yuan, Mengyao Gao, Hyunook Kim, Kinjal J Shah, Si-Lu Pei, and Pen-Chi Chiang. 2018. Advances and Challenges in Sustainable Tourism toward a Green Economy. *Science of The Total Environment* 635: 452–69. [CrossRef]

Pérez, Carlota. 2010. Technological Revolutions and Techno-Economic Paradigms. *Cambridge Journal of Economics* 34: 185–202. [CrossRef]

Provalis Research. 2015. QDA Miner (Version 5). Available online: https://provalisresearch.com (accessed on 25 May 2019).

Ronzon, Tévécia, and Robert M'Barek. 2018. Socioeconomic Indicators to Monitor the EU's Bioeconomy in Transition. *Sustainability* 10: 1745. [CrossRef]

Ruban, Dmitry A., Tatyana K. Molchanova, and Natalia N. Yashalova. 2019. Three Rising Tourism Directions and Climate Change: Conceptualizing New Opportunities. *E-Review of Tourism Research* 16: 352–70.

Ruiz-Lozano, Mercedes, Araceli De-los-Ríos-Berjillos, and Salud Millán-Lara. 2018. Spanish Hotel Chains Alignment with the Global Code of Ethics for Tourism. *Journal of Cleaner Production* 199: 205–13. [CrossRef]

Saaty, Thomas Lorie. 1996. *Decision Making with Dependence and Feedback: The Analytic Network Process*. Pittsburgh: RWS Publications, vol. 4922.

Saaty, Thomas L. 2005. *Theory and Applications of the Analytic Network Process: Decision Making with Benefits, Opportunities, Costs, and Risks*. Pittsburgh: RWS Publications.

Satchatippavarn, Sireethorn, Elias Martinez-Hernandez, Melissa Yuling Leung Pah Hang, Matthew Leach, and Aidong Yang. 2016. Urban Biorefinery for Waste Processing. *Chemical Engineering Research and Design* 107: 81–90. [CrossRef]

Schäpke, Niko, Ines Omann, Julia M. Wittmayer, Frank van Steenbergen, and Mirijam Mock. 2017. Linking Transitions to Sustainability: A Study of the Societal Effects of Transition Management. *Sustainability* 9: 737. [CrossRef]

Smith, Adrian, Andy Stirling, and Frans Berkhout. 2005. The Governance of Sustainable Socio-Technical Transitions. *Research Policy* 34: 1491–510. [CrossRef]

Starr, Morgan, Omkar Joshi, Rodney E. Will, and Chris B. Zou. 2019. Perceptions Regarding Active Management of the Cross-Timbers Forest Resources of Oklahoma, Texas, and Kansas: A SWOT-ANP Analysis. *Land Use Policy* 81: 523–30. [CrossRef]

UNEP/GPA. 2006. The State of the Marine Environment: Trends and Processes. UNEP Global Programme of Action. Available online: https://www.unenvironment.org/resources/report/state-marine-environment-trends-and-processes (accessed on 22 April 2019).

Vargas Payera, Sofía. 2018. Understanding Social Acceptance of Geothermal Energy: Case Study for Araucanía Region, Chile. *Geothermics* 72: 138–44. [CrossRef]

Wang, Yao, and Qiang Zhi. 2016. The Role of Green Finance in Environmental Protection: Two Aspects of Market Mechanism and Policies. *Energy Procedia* 104: 311–16. [CrossRef]

Wocke, Albert, and Mvd Merwe. 2007. An Investigation into Responsible Tourism Practices in the South African Hotel Industry. *South African Journal of Business Management* 38: 1–15.

Zuo, Jian, and Zhen-Yu Zhao. 2014. Green Building Research–Current Status and Future Agenda: A Review. *Renewable and Sustainable Energy Reviews* 30: 271–81. [CrossRef]

© 2019 by the author. Licensee MDPI, Basel, Switzerland. This article is an open access article distributed under the terms and conditions of the Creative Commons Attribution (CC BY) license (http://creativecommons.org/licenses/by/4.0/).

Article

Identifying the Equilibrium Point between Sustainability Goals and Circular Economy Practices in an Industry 4.0 Manufacturing Context Using Eco-Design

Fernando E. Garcia-Muiña [1], Rocío González-Sánchez [1], Anna Maria Ferrari [2], Lucrezia Volpi [2], Martina Pini [2], Cristina Siligardi [3] and Davide Settembre-Blundo [1,4,*]

1. Department of Business Administration (ADO), Applied Economics II and Fundaments of Economic Analysis, Rey-Juan-Carlos University, 28032 Madrid, Spain
2. Department of Sciences and Methods for Engineering, University of Modena and Reggio Emilia, 42122 Modena, Italy
3. Department of Engineering "Enzo Ferrari", University of Modena and Reggio Emilia, 41125 Modena, Italy
4. Project Management Office, Gruppo Ceramiche Gresmalt, 41049 Sassuolo, Italy
* Correspondence: davide.settembre@gresmalt.it; Tel.: +39-536-867-011

Received: 27 June 2019; Accepted: 12 August 2019; Published: 14 August 2019

Abstract: For manufacturing companies, the transition to circular business models (CBMs) can be hampered both by the lack of relevant data and by operational tools. Eco-design, associated with Industry 4.0 IoT (Internet of Things) technologies, can be an effective methodological approach in developing products that are consistent with the principles of the circular economy. The reason is that, in the design phase, decisions are made that can significantly influence the degree of sustainability of products during their lifecycle. Therefore, in the manufacturing environment, eco-design represents an innovative approach to include sustainability among the traditional industrial variables such as functionality, aesthetics, quality, and profit. This study aimed to test eco-design as a tool to define the equilibrium point between sustainability and circular economy in the manufacturing environment of ceramic tile production, and to demonstrate how new business opportunities can be created through evolution from a linear to a circular business model, thanks to IoT and Industry 4.0 technologies used as enabling factors. The main result of this paper was the empirical validation in a manufacturing environment of sustainability paradigms through eco-design tools and digital technologies, proposing the circular business model as an operational tool to promote the competitiveness of enterprises.

Keywords: eco-design; sustainability; circular economy (CE); circular business models (CBMs); Industry 4.0; industrial symbiosis; industrial district (ID); Italian ceramic industry

1. Introduction

Nowadays, due to the competition, companies can no longer be based only on minimizing costs, but must also use their innovative capacity to increase the environmental quality of products (Panigrahi 2017). Improving environmental performance can open up new market segments to companies that were previously unexplored. These new consumers require detailed knowledge and information about the environmental costs of what they consume and use; therefore, they are capable of enabling a product's success, one that includes the attributes of quality and design as well as sustainability—that is, a product with equal functional and aesthetic performances with as little impact as possible on the environment and society (Ceschin and Gaziulusoy 2016). Thus, companies that want to direct their innovative capacity towards the principles of sustainability will have to adopt eco-design, which is an approach in which the environmental variable assumes strategic importance

(Romli et al. 2015). In this new design approach, attention to aesthetics, functionality, and cost are integrated with assessments of the flows of energies, resources, and materials needed to manufacture and use products in order to reduce their impact on the external environment, making them sustainable also from an economic–social point of view (Lacasa et al. 2016). Sustainable product design is also the first step towards a circular economy. Eco-design considers the environmental effect that the product will have throughout its lifecycle, from production to disposal (Den Hollander et al. 2017). For this reason, it is necessary to use operational tools such as Life Cycle Assessments (LCAs), which allow for the selection of low-impact resources and technological solutions that minimize waste and favor the length of the product's lifecycle up to its disposal, so that it can be easily disassembled and recycled (Kulak et al. 2016).

This paper intends to explore the adoption of eco-design to minimize the environmental and socio-economic effects of the production of ceramic tiles, rationalizing the supply system and favoring the use of resources from local sources to reduce the incidence of transport as an element of environmental criticism. The rationalization of the formulations of ceramic bodies took place within a collaborative framework of industrial symbiosis with key suppliers and using life cycle tools (i.e., LCAs and Life Cycle Costings LCCs) to define alternative design scenarios. In addition to environmental and socio-economic sustainability, the technology was also determined by testing prototypes at the laboratory scale in order to demonstrate their industrial feasibility. The monitoring of sustainability performances during the production of the best solution obtained during the design phase will be carried out through the use of IoT technologies in an Industry 4.0 environment, which will facilitate the integration of the assessment tools with the management systems for the collection and processing of process and business data. Finally, eco-design has made it possible to update the circular business model by including strategies for creating and capturing value through the marketing of products that are more environmentally friendly.

2. Theoretical Background

Linear production systems, which currently dominate the global economy, have proven to be resource constrained and have a high environmental and social impact because they are fundamentally based on the extraction, manufacture, use, and disposal of end-of-life products (Nasir et al. 2017). Therefore, improving efficiency by reducing the use of resources and fossil fuels will not be enough to meet today's environmental challenges (Gusmerotti et al. 2019). Linear models are exposed to fluctuating prices and access to raw materials (for economic and geopolitical reasons) and contribute to environmental degradation by affecting ecosystem services fundamental to development. In contrast to this linear economy, the circular economy, an economic concept included in the framework of sustainable development, is becoming an increasingly attractive alternative (Schroeder et al. 2019). If resource consumption continues to increase as it has in recent years, by 2050 the world's population would need three times more materials and 70% more food (Crist et al. 2017). In the next twenty years alone, the need for water and energy will be 35–40% greater. This resource race will have a significant impact on Europe's economy, in which 40% of its total costs are due to the consumption of raw materials, compared to 20% for labor costs, and based on a commodities market in which there has been an annual price increase of 6% since 2000 (Lane 2017).

In order to identify concrete projects for a circular economy, we need to look at Europe, which is now the only region in the world that already has a roadmap on its table to start applying specific criteria and rules. The European Commission stresses that the circular economy will boost the European Union's (EU) competitiveness by protecting businesses against resource scarcity and price volatility (Domenech and Bahn-Walkowiak 2019). In this case, environmental protection, human health, innovation, and improved competitiveness are embraced to define what the European economy is expected to look like in the coming decades. The EU also points out that this new way of consuming and producing creates new business opportunities and locally appropriate jobs at all skill levels and, thus, generates opportunities for integration and social cohesion (Ghenţa and Matei 2018). To promote this new paradigm, the EU has launched various initiatives to address, in an integrated manner, some of the major challenges arising from the environmental and competitiveness problems of European industry. The "Roadmap to a Resource-Efficient Europe", framed in the European Commission's Europe 2020 Strategy, establishes actions to stimulate the market for secondary materials and the demand for recycled materials by offering economic incentives and developing criteria to determine when waste ceases to be waste (Barbosa et al. 2017). On the other hand, the Union's Seventh General Action Programme for the Environment 2013–2020 sets as its second priority the objective to turn the Union into a low-carbon (Sugiawan et al. 2019), resource-efficient, ecological, and competitive economy, (Breure et al. 2018) capable of mitigating climate change (Cucchiella et al. 2017; D'Adamo 2018). The other major European initiative is called "An Integrated Industrial Policy for the Globalization Era". It establishes six priority lines of action, among which is a sustainable industrial, construction, and raw materials policy that promotes, among others, the development of stable recycling markets and systems for extended producer responsibility, as a means of moving towards a circular economy (Lucchese et al. 2016).

The circular economy, according to the definition given by the Ellen MacArthur Foundation, "is a generic term to define an economy designed to be able to regenerate itself". In a circular economy, material flows are of two types: biological ones, capable of being reintegrated into the biosphere, and technical ones, destined to be revalued without entering the biosphere" (Korhonen et al. 2018). The circular economy is, therefore, a system in which all activities, starting from extraction and production, are organized in such a way that someone's waste becomes a resource for someone else (Fiksel and Lal 2018). Therefore, on the basis of this definition, the circular economic model ultimately seeks to decouple global economic development from finite resource consumption (Korhonen et al. 2018). It promotes key strategic objectives, such as generating economic growth (Busu 2019), creating jobs, and reducing environmental impacts, including carbon emissions (Suárez-Eiroa et al. 2019). With the economic model and linear development, we are depleting certain natural resources, so the circular economy proposes a new model of society that uses and optimizes materials and waste, giving them a second life (Paletta 2019). Thus, the product must be designed to be reused and recycled; that is, thanks to eco-design, the first to the last piece can be reused or recycled after the end of its useful life. With the circular economy, it is a question of how to convert what, up until now, has been considered waste into new raw materials (Caruso and Gattone 2019). In addition, it is also concerned with generating employment in the context of the so-called green economy. Therefore, the circular economy proposes a radical systemic change aimed at eco-design, economy of functionality, reuse, repair, remanufacturing, and industrial symbiosis (Baldassarre et al. 2019). This approach promotes innovation and long-term resilience and enables the development of new business models (Schroeder et al. 2019).

The implementation of the new philosophy of consumption and production based on the circular economy, requires, above all, training and knowledge of the different concepts associated with it. Eco-design is a key factor in the circular economy and consists of identifying, at the very moment a product/service is projected, all the environmental effects that can occur in each of the phases of its lifecycle, in order to try to reduce them to the minimum, without detriment to their quality and applications (Sauvé et al. 2016). Eco-design must consider the basic elements that make a product saleable, ranging from its appearance or aesthetics to its function, but unlike in the outdated linear economy, it must also assess all stages of its production and distribution chain, as well as economic and commercial aspects (Kuo et al. 2016). But to speak of eco-design as a model of complete product development, we must involve other concepts that consider their environmental and social repercussions. In the design of a product or service, we begin by defining its characteristics and processes: composition, raw materials to be used, how we will manufacture it, how we will transport it, and how we will market it. But we will also think about its usefulness and functionality, its durability, and how we will manage its useful life, especially in the final phase of the cycle (Castka and Corbett 2016).

Another concept linked to circularity is that of the functional economy, the purpose of which is to privilege the use over possession and, therefore, the provision of a service rather than the sale of a good. Compared to the linear economy, this new approach is aimed at the dematerialization of production processes seen as the only way to create value (Negrei and Istudor 2018). The functional economy wants to optimize the function of the use of goods and services, maximizing their value in the long run and minimizing the consumption of material resources and energy (Urbinati et al. 2019; Sassanelli et al. 2019).

Other more common principles of environmental management are the basis of the circular economy: reduce consumption, reuse, and recycle (the so-called three Rs of environmental management). However, if considered separately, they cannot be confused with good examples of circular economy (Ghisellini et al. 2016). Some people also prefer to use another name for this type of action: "downcycling", which can be characterized as using the remains of a product to generate others with less added value (Pires and Martinho 2019). In general, the circular economy goes beyond the relatively simple practice of recycling.

The real circular economy should establish channels of collaboration among the different companies in a supply chain to achieve more efficient results. In this regard, the concept of *industrial symbiosis* is increasingly gaining ground (Domenech et al. 2019). It is a strategy for the transfer and sharing of resources among industries in the same supply chain but belonging to different sectors, such as material waste, energy by-products, services, and capacity (Herczeg et al. 2018). Industrial symbiosis favors intermediation and innovative collaboration among companies, so that the waste produced by one of them is valued as a raw material for another (Desrochers and Szurmak 2017). The adoption and dissemination of this strategy, through appropriate instruments of relations among companies, allows to obtain significant advantages from an economic and environmental point of view, making production systems more sustainable overall (Yeo et al. 2019). The strategies of industrial symbiosis are, therefore, the basis of the effective circular economy (Zaman 2017). But for industrial symbiosis to operate, the different industrial systems present on the territory must be fully integrated, not only from the point of view of production, but also from that of waste disposal (Albino et al. 2016). One of the fundamental variables for assessing the feasibility of symbiosis from an economic point of view is the distance between the waste producer and the potential user (Marchi et al. 2017). If the cost of transporting is the same, and if their price is higher than the cost of purchasing raw materials, the circular system cannot work.

From all this, the strategic importance of the supply chain arises. In a linear economy, the supply chains are the ones that extract, use, and dispose, while in a circular economy, it is the supply chains that reduce, reuse, and recycle. In a circular economy, materials are constantly circulating in many different supply chains and never have to become waste (Bressanelli et al. 2018). The biggest logistical challenges in a circular economy are the unpredictability of the flow of materials, their low financial value, and diversity of goods properties (Batista et al. 2019). Therefore, the economic and sustainable management of the supply chain will be one of the basic capabilities of successful enterprises in a circular economy (De Angelis et al. 2018).

The circular economy's perspective is then to identify the amount of resources needed for human activities within the existing and available ones, i.e., by transforming goods that have reached the end of their useful life. Waste is considered a failure of the system and the only possible correction is to transform waste and scrap into resources (Jain et al. 2018). This innovative approach must begin with the design of the product, which must be designed to last, if possible, to be repairable, and (at the end of its lifecycle) to be broken down so that each part of it finds another use. It is precisely in the concepts of recycling, reduction, recovery, repair, and reuse, which are characteristics of the circular economy, that we can identify the link between sustainability and sustainable development (Olawumi and Chan 2018). Therefore, from a circular point of view, a system should function as a biological environment where everything is functional and everything is regenerated: the concept of waste does not exist because, in fact, waste becomes the basis for the development of other forms of life in a general framework of equilibrium. Despite this, the challenge is to identify a point of equilibrium, because the system, besides being potentially regenerative, should also be sustainable (Muñoz-Torres et al. 2018). It follows that, from a sustainable point of view, not everything that could be recycled, reduced, recovered, reopened, and reused is, in fact, sustainable in environmental, social, and economic terms. With an inverse reasoning, we can see the circular economy as a paradigm of sustainability, i.e., an innovative socio-economic approach to implementing sustainability in real life and business (Geissdoerfer et al. 2017). In a manufacturing environment, the equilibrium between the system's regenerative potential and environmental and socio-economic sustainability can be identified through eco-design.

In order to implement the principles of circular economy in business strategies, reducing dependence on increasingly scarce and expensive natural resources and turning waste into income, it is necessary to rethink or plan the value proposition and also the way in which you approach customers (Pieroni et al. 2019). But what is in practice easy to enunciate ideally, is more difficult to put into practice. Most companies, especially SMEs (small and medium-sized enterprises), are not yet ready to take advantage of the opportunities offered by the circular economy and remain firm on the more traditional model of linear growth (Tăchiciu 2018). Therefore, companies that want to enjoy the circular benefits will have to develop new business models that are not subject to the limits of linear thinking (Zucchella and Previtali 2019). These new circular business models (CBMs), in order to be able to intercept in an innovative way the value created in the supply chain, will not only have to lead the development of processes that have less impact on the environment (eco-efficiency), but will also have to take advantage of new growth opportunities to promote radically positive changes (eco-effectiveness) capable of guiding both economies and businesses towards sustainability (Heyes et al. 2018).

An important aid for companies in designing a circular business model comes from digital technologies, big data management, and artificial intelligence, because they allow for forms of collaborative innovation in supply chains (Garza-Reyes et al. 2019). Digitization allows the recording of data produced at all stages of production, marketing, management of inputs, waste, and their constant evaluation in terms of efficiency (Parida et al. 2019). Particularly in the paradigm of Industry 4.0, we can integrate information and knowledge systems based on collaborative networks. It allows a more efficient and optimized management of value chains, as well as the use of resources (Nascimento et al. 2018). The fourth industrial revolution, driven by digitization and huge volumes of data, represents the potential to leverage circular business models, where renewable resources are consumed, stocks are kept infinitely, and waste is eliminated. This is where Industry 4.0 and the circular economy meet and empower (Okorie et al. 2018). On the one hand, the disruptive technologies of the new industry operate as triggers for circular strategies. On the other hand, the circular economic model provides a purpose for Industry 4.0 and drives its development (Tseng et al. 2018).

At the end of this introductory theoretical review, we can derive several conclusions which constitute the conceptual basis of this research:

1. The circular economy represents a regenerative economic system that must maximize the creation of the value of the goods that are produced;
2. The system ensures the durability of resources through the elimination of inputs and outputs through looping of materials and components of products;
3. The lifecycle of the product is extended, and this extension also favors the connection among different value chains in the same and similar supply chains;
4. The circular economy can, thus, become a paradigm of sustainability through the use of eco-design to find the equilibrium of the system between regeneration capacity and minimization of environmental and socio-economic effects;
5. A circular business model can reduce operating costs by strengthening relations with stakeholders (suppliers, employees, customers, institutions, territory) and stimulating competitiveness.

This study seeks to fill the gaps in the literature regarding the relationship between sustainability principles and circular economy practices by addressing the following research questions:

RQ1. *Can eco-design be an effective tool to predict the equilibrium point between sustainability and circular economy?*

RQ2. *How can the circular economy create new business opportunities that combine environmental and social benefits?*

RQ3. *How can IoT and Industry 4.0 technologies be effective as enabling factors for the circular economy?*

3. Methodology

This study, based on the work of Garcia-Muiña et al. (2018), aims to operationally apply a procedure to implement the principles of environmental, social, and economic sustainability in a manufacturing environment, carrying out some of the specifications of the circular business model designed in the above paper. In the case in point, the experiment was carried out in a ceramic tile manufacturer that is among the top 10 Italian companies in the sector.

The Italian ceramic industry represents an industrial cluster of great importance both at the national and European level, as shown by the data included in the 2018 sector statistical survey published by the Italian Association of Ceramic Manufacturers (Confindustria Ceramica 2019). In 2018, the sector consisted of 137 companies with approximately 19,700 employees who produced 415 million square meters of tiles. Also, in 2018, the total turnover of Italian ceramic companies was 5.4 billion euros, of which 4.5 billion came from exports, accounting for 85% of turnover. In 2018, investments amounted to 508.2 million euros (9.4% of annual turnover), a value that has allowed the entire industry to exceed 2 billion euros in the five-year period. Among the reasons that can explain this orientation to innovation: the opportunities provided by national policies for the transition to Industry 4.0, fully taken by companies in the sector, and the recovery of competitiveness through more advanced technologies with the modernization of plants and production lines.

The diagram in Figure 1 shows the conceptual scheme of the empirical development of research and the operational procedure in relation to the research questions previously formulated. The first step is represented by the strategic phase of eco-design, i.e., the design of products that minimize their environmental effect and provide society with greater value than has been taken away from the environment, during the entire production process.

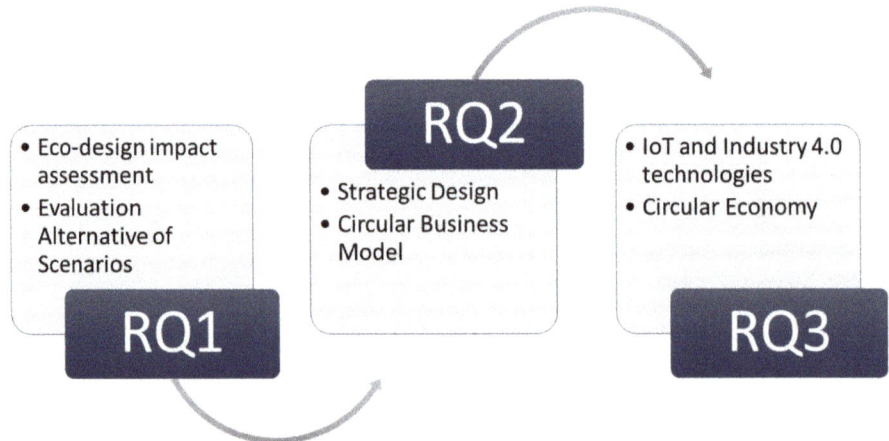

Figure 1. Conceptual diagram explaining the methodology adopted (RQ = research question; IoT = Internet of Things).

The Life Cycle Assessment (LCA) is used as a methodological tool to carry out eco-design (Eksi and Karaosmanoglu 2018). It allows for the entire lifecycle of the ceramic product to be assessed, quantifying the environmental effects from the sources of raw materials to manufacture, distribution, use, and final disposal. This is an internationally standardized procedure according to ISO 14040 and 14044. The LCA's logic is based on a holistic systemic approach that allows to understand and manage the complexity of the supply chain both upstream and downstream of the production process. Critical points in the entire product lifecycle are identified in order to envisage solutions aimed at saving and recovering energy and materials.

In order to take into account the socio-economic value of environmental damage in the tile manufacturing and industrial costs phases, the LCA analysis was supported by Life Cycle Costing (LCC) in order to predict the environmental and socio-economic sustainability of the different design scenarios (Lee et al. 2016). Like the LCA, the LCC also follows an international standard: ISO 15686. Both methodologies follow the scheme of four consequential phases, in accordance with the ISO standards: objective and scope, inventory analysis, impact assessment, and interpretation of results. Therefore, the research methodology was developed following exactly this logical scheme.

The second phase of the procedure was a strategic planning activity for a new circular business model and, finally, the third step involved the definition of the conceptual and operational links between the circular economy and sustainability.

4. Results

In a recent sustainability study carried out on a representative sample of Italian ceramic production models, it was pointed out that one of the phases of the lifecycle of the product with the greatest impact on the environment was the system of supply of raw materials (Ferrari et al. 2019). In particular, the type of transport between mines and factories (e.g., ship, train, truck) and the distance between these two locations are critical elements from an environmental point of view.

Currently, most of the raw materials used for the manufacture of ceramic tiles come from countries outside the EU—Ukraine (clays) and Turkey (feldspars). In this case, the logistics were complex; in fact, from mines, raw materials were loaded onto trucks and delivered to ports where they were shipped to Italy. Once they arrived, the materials were unloaded from ships and loaded onto trucks for transportation to factories. To a lesser extent, some clays came from Germany and, in this case, trains were used for transportation to Italy. The railway wagons arriving at the freight yard were unloaded onto trucks for delivery to factories.

Considering that about 20 kg (0.02 tons) of raw materials are needed to produce 1 square meter of tile, the Italian ceramic industry has an annual requirement of raw materials equal to:

$$415 \text{ million m}^2/\text{year} \times 0.02 \text{ tons/m}^2 = 8.3 \text{ million tons/year}$$

This figure show that the ceramic industry is a resource-intensive sector, even more so than the production process, where the transportation modes are mixed, and is a critical factor for the environment. Preliminary impact assessments (Ferrari et al. 2019) have determined that the most polluting modes of transport are ships and trucks due to their significant CO_2 emissions into the atmosphere. Trains, on the other hand, are the most ecological way of transport. Just as the distance between the source of supply and the factory is another factor that negatively affects the environmental impact.

4.1. Objective and Scope

On the assumption of this baseline, it was decided to focus the eco-design activity on optimizing the supply system to privilege more environmentally friendly transport systems, such as trains, and reducing distances between factories and mines, also using local raw materials. In order to re-engineer the ceramic material, changing the current compositional mix, it was necessary to work closely with key suppliers and other stakeholders who were represented in the same supply chain (Figure 2).

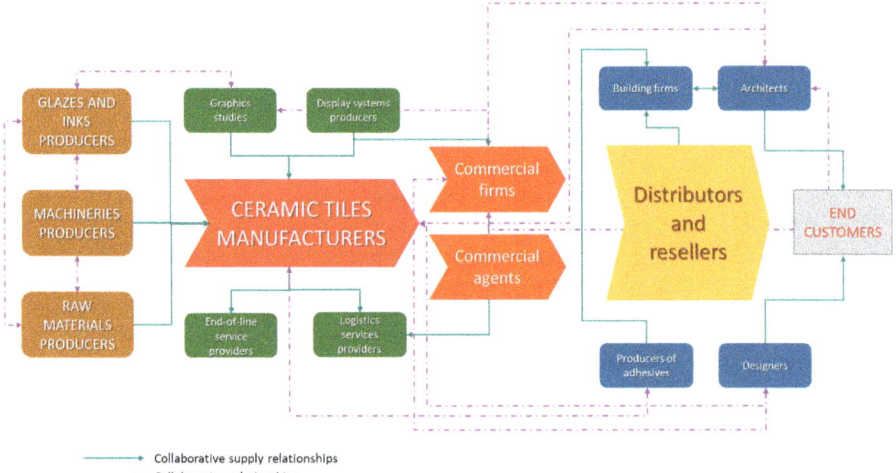

Figure 2. Ceramic supply chain network with collaborative relationships as the basis for industrial symbiosis.

At the heart of the supply chain were manufacturers of ceramic tiles, while upstream were suppliers of materials (raw materials, inks, and glazes for decoration) and technologies (machinery). The production process was also supported by a series of ancillary service providers: graphic development studios, companies that carry out additional processing and treatments on the finished product (cutting, polishing, lapping), and suppliers of display systems for the preparation of showrooms and exhibition stands. Downstream of the manufacturers, the distribution channel was made up of various economic agents: the commercial networks of the tile manufacturers, the commercial agents external to them, and the distributors. In addition, there was another category of companies, which only carried out one commercial activity, i.e., they obtained their supplies from ceramic manufacturers who manufacture the products they require under the brands of these companies.

Producers and commercial companies find themselves competing in the same markets with similar products (having shared the same technology), but mutual interest prevails: for producers to saturate production capacity by reducing industrial costs and for commercial companies to have the product to be placed on the market. From this point, the ceramic supply chain relates to the construction sector and its main economic agents: architects, designers, manufacturers of materials and solutions for the installation of floors and walls, builders, up to the end customer.

Figure 2 also shows, by means of vectors, the dynamics of collaborative relations between economic agents with and without commercial contributions for the supply of goods or services. It is clear that the supply chain is a complex system with B2B2C (business-to-business-to-consumer) characteristics, because tile manufacturers are increasingly oriented towards disintermediation of the commercial relationship by directly interacting with architects and designers overtaking distributors (Brotspies and Weinstein 2018). This relational network, typical of industrial districts, is a powerful enabling factor for industrial symbiosis and the implementation of the circular economy. Supply chain enterprises, organized in a district system, are already used to collaborate in the co-design of new technological solutions and new products. Therefore, it was decided to exploit this propensity to share knowledge to innovate the way ceramic materials are formulated, thanks to eco-design and collaboration with mining companies.

In accordance with ISO specifications for LCA and LCC analysis, 1 m² of ceramic tiles was adopted as a functional unit and the system boundaries were set from the cradle (raw materials) to the gate (end of the manufacturing process). The analysis was modelled in SimaPro®8.5.2.2 software by PRéConsultants (PRéConsultants n.d.), taking the Ecoinvent 3.4 (Wernet et al. 2016) database as a reference, especially for background processes related to natural gas, electricity, heat, transport, infrastructure, machinery, and waste treatments. The data for the impact assessment came mainly (80%) from primary sources through direct collection in the different phases of the production processes. The remaining data, on the other hand, were obtained from specialized databases.

4.2. Inventory Analysis

In order to implement an eco-design strategy, it is necessary to know the starting point in order to foresee alternative scenarios for environmental improvement. For this reason, a preliminary sustainability assessment is required, starting with an inventory analysis that defines and quantifies the input and output flows in the lifecycle of the system, building a model that represents it as truthfully as possible.

First, all the phases of the lifecycle and their relationships were displayed in a process diagram, thus determining all the inputs and outputs and, therefore, the data to be collected. This scheme is shown in Figure 3, where the main production phases of the ceramic product are represented.

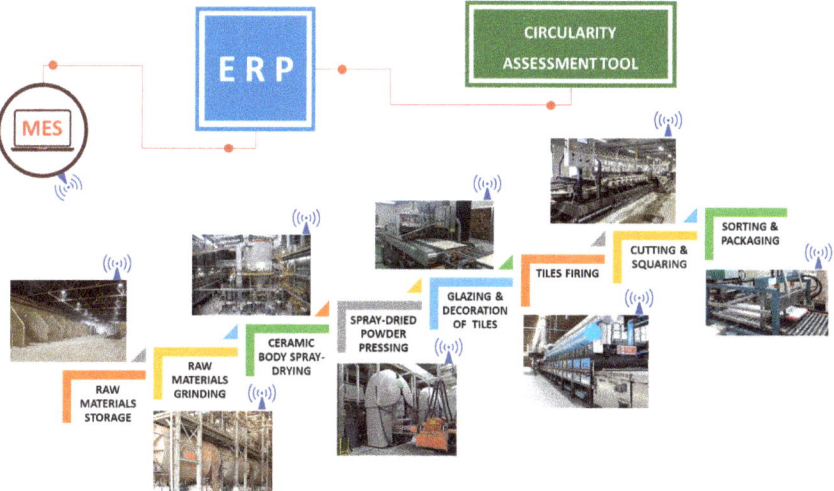

Figure 3. Ceramic production process layout with smart data collection system scheme.

The manufacturing process begins with the reception and storage of the raw materials that will be used to prepare the ceramic mixture. Changing the procurement and transport strategy in a radical way involves a different management of incoming flows and storage spaces. In this phase, collaboration with mining industries is of fundamental importance because, in a perspective of industrial symbiosis, it may be necessary to activate a sharing of the corresponding storage capacities to respond both to the criticality of transport and to the volatility of the demand for finished products.

After storage, the raw materials are mixed (with the compositions of the ceramic body of production) and ground with water in continuous rotary mills until a solid/liquid suspension called slip is obtained. This is then stored in underground tanks equipped with agitators. Special pumps take the slip and nebulize it inside a vertical dryer (spry-dryer), where the high pressure and high temperature cause the evaporation of the grinding water producing a very fine and homogeneous powder, ready to be pressed. During the pressing phase, the powders are dosed and transported to the hydraulic presses, which exert a pressure of over 490–500 kg/cm² on the spry-dried powder to form the support in the format (square or rectangular) and in the desired size. The pressed support is then covered with a layer of glaze and digitally decorated with special inks to obtain the required graphic design. At this point, the pressed, glazed, and decorated tiles are led to the kilns for firing at temperatures that reach 1210–1230 °C with cycles of 35–50 min depending on the size. The tiles coming out of the kiln can follow two paths: they can go directly to the packaging department of the finished product or they can be sent for further processing, which can include informed cutting of smaller and more modular tiles and/or lapping of the surface to obtain a brilliant effect such as stone materials (marble and granite).

For each phase of the process described above, data were collected on material flows, energy consumption (thermal and electrical), and emissions into the atmosphere. This procedure was implemented by exploiting the potential of IoT technologies, as the production plant analyzed was fully digitalized in line with the Industry 4.0 paradigm. As shown in Figure 2, smart meters were installed for each machine to monitor energy consumption in real time and to collect production data. This network of sensors was wirelessly connected with the MES (manufacturing execution system), a computer system that governs and controls the entire production process, from the release of the order to the finished product, aligning the business management needs with those of the factory and, thus, bridging the gap between the decision-making level and the executive level. The MES was then integrated with the ERP (enterprise resource planning) providing real-time data on the execution of processes to allow, in addition to the current management of operations, also the inventory analysis for environmental assessment (i.e., LCA). Since the ERP system is a common and shared database of transactional data from different sources in the organization (accounting, procurement, sales, production, and logistics), it has all the information needed to carry out the inventory analysis for the economic assessment (i.e., LCC).

4.3. Eco-Design Impact Assessment

With eco-design, we intended to evaluate the environmental and economic behavior of alternative ceramic body compositions with respect to current production, modifying the supply strategy. The new formulations are shown in Table 1.

Table 1. Alternative compositional scenarios of ceramic bodies (EU = European union; P = identification code of the compositions).

RAW MATERIALS (%)	P 01	P 03	P 04	P 15	P 17	P 19
Extra-EU clays	25	25	10	5	-	-
EU clays	25	20	45	50	28	29
Local clays	-	-	-	-	30	30
Extra-EU feldspar	38	18	19	18	19	20
Local feldspar	5	19	10	24	10	11
Local feldspar sand	7	10	11	-	10	10
Fired waste milled	-	8	5	3	3	-
Total local raw materials	12	37	26	27	53	51

The composition P 01 was the starting point, i.e., the current production. It was characterized by a wide use of imported raw materials, 63% of which came from mines located outside the European Union and transported by ship and truck over long distances (Ukraine and Turkey, 2500–3000 km). The eco-design was, therefore, focused on three goals:

1. To minimize the use of extra-EU raw materials, favoring rail over sea and road transport;
2. To valorize local raw materials for their proximity to the factory;
3. To evaluate the possibility of using the fired waste generated during manufacture, using it as a substitute for imported feldspars by exploiting their melting properties.

Therefore, the quantity of clay from outside the EU (coming from Ukraine) was progressively reduced to the advantage of a European clay that was delivered to the factory by train from Germany. In parallel, the quantity of extra-EU feldspar (coming from Turkey) was progressively replaced with a local one (Dondi et al. 2014). In addition, quantities of fired waste were introduced on a scalar basis to verify technological feasibility and environmental impact (Table 1, compositions P 03, P 04, P 15, and P 17). The extra-EU clay was then completely removed using a large quantity of local clay (composition P 17). Finally, by comparing the compositions P 17 and P 19, it was decided to verify the environmental effect of the presence or absence of fired waste with the same composition.

Based on the inventory analysis described in Section 4.2 and considering the production process shown in Figure 2 as constant, six alternative supply scenarios were simulated, corresponding to the different body compositions indicated in Table 1. For each of them, the environmental effect was determined through a predictive LCA analysis (Hauschild et al. 2018). The results of the characterization obtained with the IMPACT 2002+ assessment method is shown in Table 2, at a mid-point level (Jolliet et al. 2003).

The results highlight that the P 01 composition showed the highest effects in almost all impact categories, as clearly demonstrated by the highest value of each index. In particular, for the respiratory inorganics impact category, which refers to respiratory effects caused by inorganic substances, the impact was 31.7% higher than for composition P 17, which showed a lower impact; this was mainly caused by the emissions of nitrogen oxides in the air, especially due to the transport of raw materials by barge. Similarly, for the land occupation impact category, which takes into account the occupation of the soil, the impact related to composition P 01 was 32.1% higher than for composition P 17, primarily due to the land occupation related to the building for the extraction of the clay, for which the amount changed among the different compositions.

Moreover, for the aquatic eutrophication impact category, which refers to an abundance of nutrients in the aquatic environment, in particular nitrates and phosphates, the impact related to composition P 01 was 15.4% higher than for composition P 17, especially due to the emissions of phosphate in water caused by the treatment of sulfidic tailings coming from the manufacturing of the building for the extraction of the clay, for which the amount also varied. Finally, with regard to the global warming impact category, which considers the effects of greenhouse gases, the impact related to composition P 01 was 16.1% higher than for composition P 17, in particular due to the carbon dioxide emissions resulting from the transport by barge of raw materials.

Economic sustainability was assessed with the Life Cycle Costing (LCC) tool, which determines all the costs that a product generates during its lifecycle (Ciroth et al. 2015). The calculation was carried out in two phases (Andersson et al. 2016). In the first phase, we determined the economic costs attributable to the environmental effects generated by the product over its entire lifecycle (Table 2, above).

In this case, the economic value of externalities was determined, i.e., the environmental costs that the company should internalize in its industrial costs. This procedure is often referred to as Environmental LCC (E-LCC). The second phase, on the other hand, considered the industrial costs of the different phases of the lifecycle incurred exclusively by the company to manufacture the ceramic tiles (Table 2, below). This approach is referred to as Conventional LCC (C-LCC).

Regarding E-LCC, the calculation method EPS 2015dx (Steen 1999) determines the economic value of pollutant emissions based on the principle of the willingness to pay (WTP) by the polluter, both to remedy the damage caused and to avoid further deterioration compared to the situation created. The method identifies six main categories of damage: ecosystem services, access to water, biodiversity, building technology, human health, and abiotic resources.

Clearly, the results of the environmental impacts determined by the LCA are also reflected in the environmental externalities (E-LCC). Again, the most relevant factor in terms of external costs was the distance of the mines from the factory and the transport system used. As shown in Table 2 (above), there was a progressive decrease in externalities as the quantity of local raw materials increased and rail transport was used (from composition P 01 to composition P 19).

Table 2. Lifecycle assessment (LCA) for 1 m² of ceramic tiles.

IMPACT CATEGORIES	Unit	Alternative Scenarios					
		P 01	P 03	P 04	P 15	P 17	P 19
Carcinogens	kg C_2H_3Cl-eq	4.55×10^{-1}	4.50×10^{-1}	4.45×10^{-1}	4.45×10^{-1}	4.42×10^{-1}	4.43×10^{-1}
Non-carcinogens	kg C_2H_3Cl-eq	1.23×10^{-1}	1.21×10^{-1}	1.18×10^{-1}	1.18×10^{-1}	1.16×10^{-1}	1.17×10^{-1}
Respiratory inorganics	kg $PM_{2.5}$-eq	8.27×10^{-3}	7.47×10^{-3}	6.67×10^{-3}	6.44×10^{-3}	6.28×10^{-3}	6.36×10^{-3}
Ionizing radiation	Bq C-14-eq	3.93×10^{1}	3.66×10^{1}	3.34×10^{1}	3.27×10^{1}	3.17×10^{1}	3.21×10^{1}
Ozone layer depletion	kg CFC-11-eq	1.32×10^{-6}	1.27×10^{-6}	1.21×10^{-6}	1.19×10^{-6}	1.17×10^{-6}	1.18×10^{-6}
Respiratory organics	kg C_2H_4-eq	3.80×10^{-3}	3.58×10^{-3}	3.36×10^{-3}	3.31×10^{-3}	3.26×10^{-3}	3.29×10^{-3}
Aquatic ecotoxicity	kg TEG water	5.68×10^{2}	5.55×10^{2}	5.37×10^{2}	5.33×10^{2}	5.25×10^{2}	5.28×10^{2}
Terrestrial ecotoxicity	kg TEG soil	1.41×10^{2}	1.41×10^{2}	1.35×10^{2}	1.35×10^{2}	1.30×10^{2}	1.32×10^{2}
Terrestrial acid/nutri	kg SO_2-eq	1.78×10^{-1}	1.54×10^{-1}	1.31×10^{-1}	1.23×10^{-1}	1.20×10^{-1}	1.22×10^{-1}
Land occupation	m²org.arable	5.14×10^{-1}	4.69×10^{-1}	4.21×10^{-1}	4.00×10^{-1}	3.89×10^{-1}	3.94×10^{-1}
Aquatic acidification	kg SO_2 eq	3.32×10^{-2}	2.95×10^{-2}	2.61×10^{-2}	2.50×10^{-2}	2.44×10^{-2}	2.47×10^{-2}
Aquatic eutrophication	kg PO_4 P-lim	7.93×10^{-4}	7.51×10^{-4}	7.08×10^{-4}	7.00×10^{-4}	6.87×10^{-4}	6.93×10^{-4}
Global warming	kg CO_2-eq	7.17	6.81	6.40	6.31	6.18	6.23
Non-renewable energy	MJ primary	1.78×10^{2}	1.73×10^{2}	1.67×10^{2}	1.66×10^{2}	1.64×10^{2}	1.64×10^{2}
Mineral extraction	MJ surplus	5.27	5.11	4.95	4.89	4.88	4.89
Renewable energy	MJ	4.69	4.57	4.44	4.42	4.38	4.40
Non-carcinogens, indoor	kg C_2H_3Cl-eq	1.46×10^{-9}	1.46×10^{-9}	1.46×10^{-9}	1.46×10^{-9}	1.46×10^{-9}	1.46×10^{-9}
Respiratory organics, indoor	kg C_2H_4-eq	3.67×10^{-10}	3.67×10^{-10}	3.67×10^{-10}	3.67×10^{-10}	3.67×10^{-10}	3.67×10^{-10}
Respiratory inorganics, indoor	kg $PM_{2.5}$-eq	5.01×10^{-11}	5.01×10^{-11}	5.01×10^{-11}	5.01×10^{-11}	5.01×10^{-11}	5.01×10^{-11}
Carcinogens, indoor	kg C_2H_3Cl-eq	3.76×10^{-8}	3.76×10^{-8}	3.76×10^{-8}	3.76×10^{-8}	3.76×10^{-8}	3.76×10^{-8}
Non-carcinogens, local	kg C_2H_3Cl-eq	9.11×10^{-3}	9.11×10^{-3}	9.11×10^{-3}	9.11×10^{-3}	9.11×10^{-3}	9.11×10^{-3}
Carcinogens, local	kg C_2H_3Cl-eq	2.35×10^{-1}	2.35×10^{-1}	2.35×10^{-1}	2.35×10^{-1}	2.35×10^{-1}	2.35×10^{-1}
Respiratory organics, local	kg C_2H_4-eq	2.54×10^{-3}	2.54×10^{-3}	2.54×10^{-3}	2.54×10^{-3}	2.54×10^{-3}	2.54×10^{-3}
Respiratory inorganics, local	kg $PM_{2.5}$-eq	3.13×10^{-4}	3.13×10^{-4}	3.13×10^{-4}	3.13×10^{-4}	3.13×10^{-4}	3.13×10^{-4}

The structure of industrial costs remained broadly unchanged in the different compositions, except for the cost of raw materials, as shown in Table 3 below. The item "cost of raw materials" includes the cost of the different materials used and the corresponding transport cost. Therefore, the greater the distance between the source of supply and the factory, the greater the cost. Furthermore, for compositions containing fired waste (P 03, P 04, P 15, and P 17), their additional grinding costs must be considered. In fact, in order to be able to use them as a partial replacement for a raw material, it is necessary to reduce their size in powder form, because they are particularly hard materials and, therefore, inadequate for reintroduction as they are into the process.

Table 3. Life Cycle Costing (LCC) for 1 m² of ceramic tile.

		ENVIRONMENTAL LCC (E-LCC)					
DAMAGE CATEGORIES	Unit	Alternative Scenarios					
		P 01	P 03	P 04	P 15	P 17	P 19
Ecosystem services	€/m²	3.19×10^{-2}	3.02×10^{-2}	2.84×10^{-2}	2.80×10^{-2}	2.74×10^{-2}	2.76×10^{-2}
Access to water	€/m²	1.88×10^{-3}	1.79×10^{-3}	1.69×10^{-3}	1.67×10^{-3}	1.64×10^{-3}	1.65×10^{-3}
Biodiversity	€/m²	1.03×10^{-4}	9.75×10^{-5}	9.17×10^{-5}	9.03×10^{-5}	8.85×10^{-5}	8.92×10^{-5}
Building technology	€/m²	2.80×10^{-4}	2.66×10^{-4}	2.52×10^{-4}	2.49×10^{-4}	2.44×10^{-4}	2.46×10^{-4}
Human health	€/m²	1.32	1.25	1.17	1.15	1.13	1.14
Abiotic resources	€/m²	3.14	3.07	2.98	2.98	2.92	2.94
TOTAL (€/m²)		4.49	4.35	4.18	4.16	4.08	4.11
		CONVENTIONAL LCC (C-LCC)					
COST ITEMS	Unit	Alternative Scenarios					
		P 01	P 03	P 04	P 15	P 17	P 19
Raw materials	€/m²	1.81	1.77	1.64	1.53	1.37	1.29
Electrical energy	€/m²	0.34	0.34	0.34	0.34	0.34	0.34
Thermal energy	€/m²	0.57	0.57	0.57	0.57	0.57	0.57
Consumables	€/m²	0.75	0.75	0.75	0.75	0.75	0.75
Packages	€/m²	0.28	0.28	0.28	0.28	0.28	0.28
Human resources	€/m²	1.45	1.45	1.45	1.45	1.45	1.45
Accessories	€/m²	1.09	1.09	1.09	1.09	1.09	1.09
Amortizations	€/m²	0.56	0.56	0.56	0.56	0.56	0.56
TOTAL (€/m²)		6.85	6.81	6.68	6.57	6.41	6.33

In this study, the social dimension of sustainability was determined through the approach of the Societal Life Cycle Costing (S-LCC), which prescribes the sum of environmental externalities (E-LCC) with industrial costs (C-LCC) (De Menna et al. 2018). This approach, compared to other methodologies such as S-LCA (Petti et al. 2018), allows to directly correlate a social indicator to the functional unit, which, in our case, corresponded to 1 m² of ceramic tile (Table 4). It is particularly important to maintain the focus of the analysis on the manufacturing process by following the cycles and times in a more dynamic time horizon than the classic social analysis, as required by the guidelines (UNEP/SETAC 2009) of United Nations Environment Programme (UNEP) and the Society of Environmental Toxicology & Chemistry (SETAC).

Table 4. Societal LCC for 1 m² of ceramic tiles.

	P 01	P 03	P 04	P 15	P 17	P 19
Environmental LCC (€/m²)	4.49	4.35	4.18	4.16	4.08	4.11
Conventional LCC (€/m²)	6.85	6.81	6.68	6.57	6.41	6.33
Societal LCC (€/m²)	11.34	11.16	10.86	10.73	10.49	10.44

The S-LCC showed that the use of local raw materials and more environmentally friendly transport systems such as rail have a significant and positive socio-economic impact, especially when comparing the extreme compositions P 01 and P 19. In addition, it was also shown that recycling of processing waste was not as effective in mitigating impacts as were variations in body compositions. The sustainability assessment indicated that P 17 and P 19 were the most performing compositions from an environmental point of view, with almost equivalent impact results. However, the economic analysis showed that recycling of fired waste was not beneficial for the higher incidence of pre-grinding costs. Therefore, the best result was the composition P 19.

4.4. Interpretation and Discussion of the Results

Table 5 shows the main indicators of environmental, socio-economic, and technological sustainability, with eco-design for the various ceramic body compositions. In particular, the best result obtained in comparison with the starting point is highlighted.

Table 5. Overview of environmental, socio-economic, and technological sustainability indicators.

SUSTAINABILITY INDICATORS	P 01 Starting Point	P 03	P 04	P 15	P 17	P 19 Best Solution
Local raw materials (%)	12	37	26	27	53	51
Fired waste milled (%)		8	5	3	3	
ENVIRONMENTAL SUSTAINABILITY						
Respiratory inorganics (kg $PM_{2.5\text{-eq}}$)	8.27×10^{-3}	7.47×10^{-3}	6.67×10^{-3}	6.44×10^{-3}	6.28×10^{-3}	6.36×10^{-3}
Land occupation (m^2org.arable)	5.14×10^{-1}	4.69×10^{-1}	4.21×10^{-1}	4.00×10^{-1}	3.89×10^{-1}	3.94×10^{-1}
Aquatic eutrophication (kg PO_4 P-lim)	7.93×10^{-4}	7.51×10^{-4}	7.08×10^{-4}	7.00×10^{-4}	6.87×10^{-4}	6.93×10^{-4}
Global warming (kg $CO_{2\text{-eq}}$)	7.17	6.81	6.40		6.31	6.18	6.23
SOCIO-ECONOMIC SUSTAINABILITY						
Environmental LCC (€/m^2)	4.49	4.35	4.18	4.16	4.08	4.11
Conventional LCC (€/m^2)	6.85	6.81	6.68	6.57	6.41	6.33
Societal LCC (€/m^2)	11.34	11.16	10.86	10.73	10.49	10.44
TECHNOLOGICAL SUSTAINABILITY						
Dimensional quality (ISO 10545-2)	Conform	Not Conform	Not Conform	Not Conform	Conform	Conform
Water absorption (ISO 10545-3)	Conform	Not Conform	Conform	Not Conform	Conform	Conform
Bending strength (ISO 10545-4)	Conform	Not Conform	Conform	Not Conform	Conform	Conform

The LCA study confirmed that the distance of the sources of supply from the factory and the type of transport used were potentially critical variables for the effects that they can generate on the environment. The scenarios defined with eco-design have indicated possible alternatives to the composition of the current ceramic body, which are more respectful of the environment.

Compositional changes showed that it is possible to significantly reduce emissions into the atmosphere of particulate matter resulting from the combustion of fossil fuels that emit aerosols, sulphates, and nitrates and that can cause respiratory difficulties (impact category: respiratory inorganics). Emissions of nitrogen-containing pollutants into the environment also contribute to eutrophication, i.e., an overabundance of nitrates in water systems. This causes algae blooms that deplete the oxygen dissolved in water, consequently suffocating aquatic life (impact category: aquatic eutrophication). Similarly, new compositions of ceramic bodies can reduce the amount of carbon dioxide (CO_2) released into the atmosphere due to the combustion of fossil fuels related to truck and ship transport systems. Carbon dioxide and other greenhouse gases accumulate in the atmosphere, trapping solar heat which, in turn, increases the average temperature of the Earth causing the retreat of the glaciers, the extinction of species, the loss of soil moisture, and more extreme weather conditions (impact category: global warming). A further benefit was also obtained in the category of land occupation damage, thanks to the significant use of local raw materials that use less complex mining facilities.

Furthermore, by modifying the composition of the ceramic bodies and the transport mix to maximize the use of local raw materials, reducing the distances between mines and the factory and by favoring rail transport, it was possible to estimate a reduction in externalities (E-LCC) of about 9%, comparing the initial composition (P 01) with the best result obtained (P 19). Regarding industrial costs (C-LCC), it can be noted that the introduction of fired waste as a substitute for extra-EU feldspar was not economically viable, because the externality benefit was offset by the cost of grinding the waste. Comparing the P 01 and P 19 compositions, the advantage in terms of industrial costs was always 9%, with a comparable benefit (9%) also for societal costs (S-LCC).

However, eco-design must not only be limited to the prediction of the environmental and socio-economic performance of alternative compositional scenarios but must also assess the technical and industrial feasibility of these options. We could, therefore, speak of technological sustainability to indicate that a solution complies with a set of internal specifications and/or international quality standards. For this very reason, the compositions designed were tested at the laboratory level to verify their compliance with three international standards in force for ceramic tiles. In particular, the value of water absorption (ISO 10545-3) that determines the degree of porosity of the ceramic product, the dimensions (ISO 10545-2) that establish the geometric conformity of the tiles, and the resistance to bending, which measures their mechanical properties (ISO 10545-4). The results of this further evaluation are shown in Table 5, from which it can be seen that compositional solutions with better sustainability performance than the starting point are not always manufacturable in compliance with the regulations in force.

5. Conclusions

With this paper, a managerial example of the introduction of the circular economy paradigm in business operations was provided. In particular, it intended to redesign the business model of a ceramic tile manufacturer through the approach of eco-design in order to optimize the supply system of raw materials to improve the environmental and socio-economic performance of the finished product.

Eco-design served to provide alternative compositional scenarios demonstrating how much the distance of the sources of supply from the factory and the transport systems used can affect the environmental and socio-economic performance of the company. With the same design approach, a feasibility study was also carried out to recycle the fired waste generated during the production process. The assessment showed that it is possible to optimize the compositions in conjunction with the company objectives in terms of sustainability. The study made it possible to identify a composition of ceramic body that performed particularly well from an environmental and socio-economic point of view, compared to the current production. This result was achieved thanks to a radical change in the composition: raw materials from outside the EU were replaced by others from local mines. This made it possible to reduce the negative impact of road transport on the environment.

The assessment also showed that recycling waste was not always beneficial from a sustainability perspective. In fact, the cost of grinding the waste baked to be used in the manufacture of tiles offset the benefit of lower external costs obtained through the recycling of waste. This result can only be considered as apparently negative. In fact, it shows that the adoption of the circular economy paradigm requires a rigorous management approach, technical skills and effective tools to quantify the effects. Only with these premises can the redefinition of the business model be effective because it will have the support of a strong scientific basis.

These results provided a positive answer to the QR1 research question: eco-design is, therefore, an effective tool to predict the equilibrium point between sustainability and circular economy.

The predictive sustainability assessment was then validated at the laboratory scale in order to verify whether the new compositions conformed to the technical specifications established by the international standards for ceramic tiles. The aim was to identify this technical conformity as technological sustainability, a further dimension of sustainability which, alongside the environment, economy, and society, aims to demonstrate that the design scenario is industrially feasible. Thanks to eco-design, it was, therefore, possible to innovate the way raw materials are supplied and the industrial symbiosis within the supply chain made it possible to rationalize and make a new ceramic composition feasible. It will be the basis for the development of a finished product with better performance from the point of view of sustainability, capable of satisfying the demand for greener building products.

The development conducted in this research led to an update of the circular business model already defined previously (Garcia-Muiña et al. 2018); the changes are shown in the diagram in Table 6 using the business model canvas (Joyce and Paquin 2016). As a value proposition, the integration between IoT technologies and sustainability monitoring systems was highlighted in order to better exploit local resources and to innovate organizational models. Networking activities were added to the key activities, performed mainly at the level of the industrial district in a framework of industrial symbiosis (Morales et al. 2019; Fraccascia et al. 2016) in order to involve raw material suppliers in a cooperative way (who play both the role of key partners and key stakeholders) in the production of products that are more environmentally friendly thanks to the use of efficient and digitized production units. These products are aimed at new market segments, such as green consumers, architects, and designers, who are more sensitive to the socially responsible behavior of the industry, also using innovative distribution channels such as digital ones. The higher costs incurred in internalizing environmental and social externalities will be offset by lower production costs and an improved reputation among stakeholders. These conclusions answer the *QR2* research question: the way to create new business opportunities by intercepting the value they generate is to prepare a circular business model that replaces the linear one (Antikainen et al. 2015).

Another important result of this study was the demonstration of the effectiveness of the Industry 4.0 paradigm as an enabling factor for sustainability, thus satisfying the *QR3* research question. In fact, thanks to the complete digitization of manufacturing, environmental, socio-economic, and technological monitoring can be carried out dynamically and in real time. On the contrary, in a non-digitized environment, sustainability assessments are conducted retrospectively based on historical datasets. There were, therefore, two innovative aspects:

1. Eco-design, in a simulation environment, allows to predict the environmental, socio-economic, and technological performance of alternative industrial solutions;
2. IoT technologies, in an Industry 4.0 environment, allow real-time measurement of effects as they occur, providing the capability to intervene on processes to mitigate them.

The predictive function of eco-design and the dynamic potential of digital assessment transform conventional sustainability analyses from purely technical activities to effective strategies of corporate social responsibility because the managerial perspective is changed from short to long term. The joint use of both these good practices offers the opportunity for decision-makers in manufacturing companies to apply, in a real and effective way, the principles of the circular economy by redesigning the business model and changing the way in which value is created and intercepted.

The high level of complexity reached by industrial systems requires increasingly transversal skills for an even more accurate understanding of reality from both a technological and social point of view. The advent of the fourth industrial revolution and the diffusion of digital technologies have led to the end of a world made up of silos of skills that struggled to integrate with other universes. To stimulate innovation, a multidisciplinary approach is fundamental, as was demonstrated in this research. Integrating the socio-economic dimension of sustainability with the environmental dimension required a contamination of knowledge: materials sciences, chemistry, process engineering, information technology, business organization, and management.

Table 6. Representation of the updated circular business model inspired by the business model canvas.

CIRCULAR BUSINESS MODEL

KEY PARTNERSHIPS	KEY ACTIVITIES	VALUE PROPOSITION	CUSTOMER RELATIONSHIPS	CUSTOMER SEGMENTS
Raw material suppliers Suppliers of glazes and inks Plant and machinery suppliers Suppliers of electricity Suppliers of methane Packaging suppliers Suppliers of chemical additives IT Solution Providers Financial services providers	Ceramic tile designs Manufacturing of ceramic tiles Marketing and sales Facilities operations & maintenance Sourcing Logistics planning Management Accounting & Control Industrial Symbiosis Networking	Provide collections of porcelain stoneware tiles totally made in Italy and with the best value for money Apply eco-design techniques to the development of new products, using ecofriendly and resource saving raw materials	Extensive sales network 1to1 interaction with distributors Offer of ancillary services to the product On-demand product development	Residential customers Commercial buildings Public buildings Business customer Green consumers Architects and Designers
KEY STAKEHOLDERS	KEY RESOURCES	To develop digital solutions for our manufacturing processes able to monitor in real time the environmental, socio-economic and technological performances To technologically valorize the local natural resources Be ready to innovate organizational models	DISTRIBUTION CHANNELS	
Private business Trade channel operators Suppliers Staff person Final consumers Competitors Public Institutions Environment Partners Trade unions Public and private organizations Media	Three manufacturing units Five logistics warehouses IT Infrastructure Human capital Operational know-how Financial assets 4.0 energy and resource-efficient factories		Large-scale retails Independent distributors Specialized stores Cloud based interactive multi-channel	

COSTS STRUCTURE	REVENUE STREAM
Manufacturing costs Commercial costs Research and development costs General and administrative costs Financing cost Environmental costs (externalities) Social costs	Volume of sales Value recovered from the use of less expensive local raw materials Better reputation from stakeholders

The search for an equilibrium between the degree of sustainability of alternative compositional scenarios and the corresponding potential for circularity in tile manufacturing required the definition of a decision area with multiple criteria, where several functions needed to be optimized at the same time (Caruso et al. 2017). In this study, the search for efficiency in the use of natural resources was pursued, both as a private and collective goal. The fundamental role of managerial sciences was, thus, demonstrated by examining the positive (or interpretative) dimension of environmental and socio-economic problems. With this contribution of knowledge and by supporting the engineering sciences, mainly focused on the regulatory aspects of the problems, it was possible to obtain a more exhaustive framework linking sustainability and circular economy in businesses to support decision-making processes.

Author Contributions: Supervision, F.E.G.-M.; Conceptualization, R.G.-S.; Data curation, A.M.F.; Formal analysis, L.V.; Methodology, M.P.; Data curation, C.S.; Writing—review & editing, D.S.-B.

Funding: This research was co-funded by the European Union under the LIFE Programme (LIFE16 ENV/IT/000307: LIFE Force of the Future-Forture).

Acknowledgments: The authors thank the editor and three anonymous reviewers for their helpful comments on this paper.

Conflicts of Interest: The authors declare no conflict of interest.

References

Albino, Vito, Luca Fraccascia, and Ilaria Giannoccaro. 2016. Exploring the role of contracts to support the emergence of self-organized industrial symbiosis networks: An agent-based simulation study. *Journal of Cleaner Production* 112: 4353–66. [CrossRef]

Andersson, Karin, Selma Brynolf, Hanna Landquist, and Erik Svensson. 2016. Methods and Tools for Environmental Assessment. In *Shipping and the Environment*. Berlin and Heidelberg: Springer, pp. 265–93.

Antikainen, Maria, Minna Lammi, Harri Paloheimo, Timo Rüppel, and Katri Valkokari. 2015. Towards circular economy business models: Consumer acceptance of novel services. Paper presented at the ISPIM Innovation Summit, Brisbane, Australia, December 6–9.

Baldassarre, Brian, Micky Schepers, Nancy Bocken, Eefje Cuppen, Gijsbert Korevaar, and Giulia Calabretta. 2019. Industrial Symbiosis: Towards a design process for eco-industrial clusters by integrating Circular Economy and Industrial Ecology perspectives. *Journal of Cleaner Production* 216: 446–60. [CrossRef]

Barbosa, Ana, Sara Vallecillo, Claudia Baranzelli, Chris Jacobs-Crisioni, Filipe Batista e Silva, Carolina Perpiña-Castillo, Carlo Lavalle, and Joachim Maes. 2017. Modelling built-up land take in Europe to 2020: An assessment of the Resource Efficiency Roadmap measure on land. *Journal of Environmental Planning and Management* 60: 1439–63. [CrossRef]

Batista, Luciano, Michael Bourlakis, Palie Smart, and Roger Maull. 2019. Business Models in the Circular Economy and the Enabling Role of Circular Supply Chains. In *Operations Management and Sustainability*. Cham: Palgrave Macmillan, pp. 105–34.

Bressanelli, Gianmarco, Marco Perona, and Nicola Saccani. 2018. Challenges in supply chain redesign for the Circular Economy: A literature review and a multiple case study. *International Journal of Production Research*, 1–28. [CrossRef]

Breure, A. M., J. P. A. Lijzen, and L. Maring. 2018. Soil and land management in a circular economy. *Science of the Total Environment* 624: 1125–30. [CrossRef]

Brotspies, Herbert, and Art Weinstein. 2018. Rethinking Business Segmentation: A Conceptual Model and Strategic Insights. *Journal of Strategic Marketing* 26. [CrossRef]

Busu, Mihail. 2019. Adopting Circular Economy at the European Union Level and Its Impact on Economic Growth. *Social Sciences* 8: 1125–30. [CrossRef]

Caruso, Giulia, and Stefano Antonio Gattone. 2019. Waste Management Analysis in Developing Countries through Unsupervised Classification of Mixed Data. *Social Sciences* 8: 186. [CrossRef]

Caruso, Giulia, Stefano Antonio Gattone, Francesca Fortuna, and Tonio Di Battista. 2017. Cluster analysis as a decision-making tool: A methodological review. In *International Symposium on Distributed Computing and Artificial Intelligence*. Cham: Springer, pp. 48–55.

Castka, Pavel, and Charles Corbett. 2016. Adoption and diffusion of environmental and social standards: The effect of stringency, governance, and media coverage. *International Journal of Operations & Production Management* 36: 1504–29.

Ceschin, Fabrizio, and Idil Gaziulusoy. 2016. Evolution of design for sustainability: From product design to design for system innovations and transitions. *Design Studies* 47: 118–63. [CrossRef]

Ciroth, Andreas, Jutta Hildenbrand, and Bengt Steen. 2015. Life Cycle Costing. In *Sustainability Assessment of Renewables-Based Products: Methods and Case Studies*. Chichester: John Wiley & Sons Ltd., pp. 215–28.

Confindustria Ceramica. 2019. *Indagine Statistica Nazionale*, 39th ed. Ceramic Tiles 2018. Sassuolo: Confindustria Ceramica Study Centre.

Crist, Eileen, Camilo Mora, and Robert Engelman. 2017. The interaction of human population, food production, and biodiversity protection. *Science* 356: 260–64. [CrossRef]

Cucchiella, Federica, Idiano D'Adamo, Massimo Gastaldi, S. C. Lenny Koh, and Paolo Rosa. 2017. A comparison of environmental and energetic performance of European countries: A sustainability index. *Renewable and Sustainable Energy Reviews* 78: 401–13. [CrossRef]

D'Adamo, Idiano. 2018. The profitability of residential photovoltaic systems. A new scheme of subsidies based on the price of CO_2 in a developed PV market. *Social Sciences* 7: 148. [CrossRef]

De Angelis, Roberta, Mickey Howard, and Joe Miemczyk. 2018. Supply chain management and the circular economy: Towards the circular supply chain. *Production Planning & Control* 29: 425–37.

De Menna, Fabio, Jana Dietershagen, Marion Loubiere, and Matteo Vittuari. 2018. Life cycle costing of food waste: A review of methodological approaches. *Waste Management* 73: 1–13. [CrossRef]

Den Hollander, Marcel C., Conny A. Bakker, and Erik Jan Hultink. 2017. Product design in a circular economy: Development of a typology of key concepts and terms. *Journal of Industrial Ecology* 21: 517–25. [CrossRef]

Desrochers, Pierre, and Joanna Szurmak. 2017. Long distance trade, locational dynamics and by-product development: Insights from the history of the American cottonseed industry. *Sustainability* 9: 579. [CrossRef]

Domenech, Teresa, and Bettina Bahn-Walkowiak. 2019. Transition towards a resource efficient circular economy in Europe: Policy lessons from the EU and the member states. *Ecological Economics* 155: 7–19. [CrossRef]

Domenech, Teresa, Raimund Bleischwitz, Asel Doranova, Dimitris Panayotopoulos, and Laura Roman. 2019. Mapping Industrial Symbiosis Development in Europe typologies of networks, characteristics, performance and contribution to the Circular Economy. *Resources, Conservation and Recycling* 141: 76–98. [CrossRef]

Dondi, Michele, Mariarosa Raimondo, and Chiara Zanelli. 2014. Clays and bodies for ceramic tiles: Reappraisal and technological classification. *Applied Clay Science* 96: 91–109. [CrossRef]

Eksi, Guner, and Filiz Karaosmanoglu. 2018. Life cycle assessment of combined bioheat and biopower production: An eco-design approach. *Journal of Cleaner Production* 197: 264–79. [CrossRef]

Ferrari, Anna, Lucrezia Volpi, Martina Pini, Cristina Siligardi, Fernando García-Muiña, and Davide Settembre-Blundo. 2019. Building a Sustainability Benchmarking Framework of Ceramic Tiles Based on Life Cycle Sustainability Assessment (LCSA). *Resources* 8: 11. [CrossRef]

Fiksel, Joseph, and Rattan Lal. 2018. Transforming waste into resources for the Indian economy. *Environmental Development* 26: 123–28. [CrossRef]

Fraccascia, Luca, Maurizio Magno, and Vito Albino. 2016. Business models for industrial symbiosis: A guide for firms. *Procedia Environmental Science, Engineering and Management* 3: 83–93.

Garcia-Muiña, Fernando E., Rocío González-Sánchez, Anna Maria Ferrari, and Davide Settembre-Blundo. 2018. The Paradigms of Industry 4.0 and Circular Economy as Enabling Drivers for the Competitiveness of Businesses and Territories: The Case of an Italian Ceramic Tiles Manufacturing Company. *Social Sciences* 7: 255. [CrossRef]

Garza-Reyes, Jose Arturo, Vikas Kumar, Luciano Batista, Anass Cherrafi, and Luis Rocha-Lona. 2019. From linear to circular manufacturing business models. *Journal of Manufacturing Technology Management* 30: 554–60. [CrossRef]

Geissdoerfer, Martin, Paulo Savaget, Nancy M. P. Bocken, and Erik Jan Hultink. 2017. The Circular Economy—A new sustainability paradigm? *Journal of Cleaner Production* 143: 757–68. [CrossRef]

Ghenţa, Mihaela, and Aniela Matei. 2018. SMEs and the Circular Economy: From Policy to Difficulties Encountered During Implementation. *Amfiteatru Economic* 20: 294–309. [CrossRef]

Ghisellini, Patrizia, Catia Cialani, and Sergio Ulgiati. 2016. A review on circular economy: The expected transition to a balanced interplay of environmental and economic systems. *Journal of Cleaner Production* 114: 11–32. [CrossRef]

Gusmerotti, Natalia Marzia, Francesco Testa, Filippo Corsini, Gaia Pretner, and Fabio Iraldo. 2019. Drivers and approaches to the circular economy in manufacturing firms. *Journal of Cleaner Production* 230: 314–27. [CrossRef]

Hauschild, Michael Z., Ralph K. Rosenbaum, and Stig Irvin Olsen. 2018. *Life Cycle Assessment*. Berlin: Springer.

Herczeg, Gábor, Renzo Akkerman, and Michael Zwicky Hauschild. 2018. Supply chain collaboration in industrial symbiosis networks. *Journal of Cleaner Production* 171: 1058–67. [CrossRef]

Heyes, Graeme, Maria Sharmina, Joan Manuel F. Mendoza, Alejandro Gallego-Schmid, and Adisa Azapagic. 2018. Developing and implementing circular economy business models in service-oriented technology companies. *Journal of Cleaner Production* 177: 621–32. [CrossRef]

Jain, Sourabh, Nikunj Kumar Jain, and Bhimaraya Metri. 2018. Strategic framework towards measuring a circular supply chain management. *Benchmarking: An International Journal* 25: 3238–52. [CrossRef]

Jolliet, Olivier, Manuele Margni, Raphaël Charles, Sébastien Humbert, Jérôme Payet, Gerald Rebitzer, and Ralph Rosenbaum. 2003. IMPACT 2002+: A new life cycle impact assessment methodology. *The International Journal of Life Cycle Assessment* 8: 324. [CrossRef]

Joyce, Alexandre, and Raymond L. Paquin. 2016. The triple layered business model canvas: A tool to design more sustainable business models. *Journal of Cleaner Production* 135: 1474–86. [CrossRef]

Korhonen, Jouni, Antero Honkasalo, and Jyri Seppälä. 2018. Circular economy: The concept and its limitations. *Ecological Economics* 143: 37–46. [CrossRef]

Kulak, Michal, Thomas Nemecek, Emmanuel Frossard, and Gérard Gaillard. 2016. Eco-efficiency improvement by using integrative design and life cycle assessment. The case study of alternative bread supply chains in France. *Journal of Cleaner Production* 112: 2452–61. [CrossRef]

Kuo, Tsai-Chi, Shana Smith, Gregory C. Smith, and Samuel H. Huang. 2016. A predictive product attribute driven eco-design process using depth-first search. *Journal of Cleaner Production* 112: 3201–10. [CrossRef]

Lacasa, Enrique, Jose Luis Santolaya, and Anna Biedermann. 2016. Obtaining sustainable production from the product design analysis. *Journal of Cleaner Production* 139: 706–16. [CrossRef]

Lane, Alexander. 2017. Business growth needs sustainable water treatments. *Filtration+ Separation* 54: 22–24. [CrossRef]

Lee, Cheul-Kyu, Jae-Young Lee, Yo-Han Choi, and Kun-Mo Lee. 2016. Application of the integrated ecodesign method using the GHG emission as a single indicator and its GHG recyclability. *Journal of Cleaner Production* 112: 1692–99. [CrossRef]

Lucchese, Matteo, Leopoldo Nascia, and Mario Pianta. 2016. Industrial policy and technology in Italy. *Economia e Politica Industriale* 43: 233–60. [CrossRef]

Marchi, Beatrice, Simone Zanoni, and Lucio E. Zavanella. 2017. Symbiosis between industrial systems, utilities and public service facilities for boosting energy and resource efficiency. *Energy Procedia* 128: 544–50. [CrossRef]

Morales, E. Manuel, Arnaud Diemer, Gemma Cervantes, and Graciela Carrillo-González. 2019. "By-product synergy" changes in the industrial symbiosis dynamics at the Altamira-Tampico industrial corridor: 20 Years of industrial ecology in Mexico. *Resources, Conservation and Recycling* 140: 235–45. [CrossRef]

Muñoz-Torres, María Jesús, María Ángeles Fernández-Izquierdo, Juana M. Rivera-Lirio, Idoya Ferrero-Ferrero, Elena Escrig-Olmedo, José Vicente Gisbert-Navarro, and María Chiara Marullo. 2018. An assessment tool to integrate sustainability principles into the global supply chain. *Sustainability* 10: 535. [CrossRef]

Nascimento, Daniel Luiz Mattos, Viviam Alencastro, Osvaldo Luiz Gonçalves Quelhas, Rodrigo Goyannes Gusmão Caiado, Jose Arturo Garza-Reyes, Luis Rocha-Lona, and Guilherme Tortorella. 2018. Exploring Industry 4.0 technologies to enable circular economy practices in a manufacturing context: A business model proposal. *Journal of Manufacturing Technology Management* 30: 607–27. [CrossRef]

Nasir, Mohammed Haneef Abdul, Andrea Genovese, Adolf A. Acquaye, S. C. L. Koh, and Fred Yamoah. 2017. Comparing linear and circular supply chains: A case study from the construction industry. *International Journal of Production Economics* 183: 443–57. [CrossRef]

Negrei, Costel, and Nicolae Istudor. 2018. Circular Economy—Between Theory and Practice. *Amfiteatru Economic* 20: 498–509. [CrossRef]

Okorie, Okechukwu, Konstantinos Salonitis, Fiona Charnley, Mariale Moreno, Christopher Turner, and Ashutosh Tiwari. 2018. Digitisation and the circular economy: A review of current research and future trends. *Energies* 11: 3009. [CrossRef]

Olawumi, Timothy O., and Daniel W. M. Chan. 2018. A scientometric review of global research on sustainability and sustainable development. *Journal of Cleaner Production* 183: 231–50. [CrossRef]

Paletta, Angelo. 2019. Rethinking Economics in a Circular Way in the Light of Encyclical "Laudato Sì". In *Sustainability and the Humanities*. Cham: Springer, pp. 339–57.

Panigrahi, Ramakrushna. 2017. Incidence of Green Accounting on Competitiveness: Empirical Evidences from Mining and Quarrying Sector. In *Business Analytics and Cyber Security Management in Organizations*. Hershey: IGI Global, pp. 270–78.

Parida, Vinit, David Sjödin, and Wiebke Reim. 2019. Reviewing Literature on Digitalization, Business Model Innovation, and Sustainable Industry: Past Achievements and Future Promises. *Sustainability* 11: 391. [CrossRef]

Petti, Luigia, Monica Serreli, and Silvia Di Cesare. 2018. Systematic literature review in social life cycle assessment. *The International Journal of Life Cycle Assessment* 23: 422–31. [CrossRef]

Pieroni, Marina P., Tim McAloone, and Daniela A. C. Pigosso. 2019. Business model innovation for circular economy and sustainability: A review of approaches. *Journal of Cleaner Production* 215: 198–216. [CrossRef]

Pires, Ana, and Graça Martinho. 2019. Waste hierarchy index for circular economy in waste management. *Waste Management* 95: 298–305. [CrossRef]

PRéConsultants. SimaPro 8.5.2.2 Multiuser. n.d. Available online: https://www.pre-sustainability.com/simapro (accessed on 27 June 2019).

Romli, Awanis, Paul Prickett, Rossitza Setchi, and Shwe Soe. 2015. Integrated eco-design decision-making for sustainable product development. *International Journal of Production Research* 53: 549–71. [CrossRef]

Sassanelli, Claudio, Paolo Rosa, Roberto Rocca, and Sergio Terzi. 2019. Circular Economy performance assessment methods: A systematic literature review. *Journal of Cleaner Production* 229: 440–53. [CrossRef]

Sauvé, Sébastien, Sophie Bernard, and Pamela Sloan. 2016. Environmental sciences, sustainable development and circular economy: Alternative concepts for trans-disciplinary research. *Environmental Development* 17: 48–56. [CrossRef]

Schroeder, Patrick, Kartika Anggraeni, and Uwe Weber. 2019. The relevance of circular economy practices to the sustainable development goals. *Journal of Industrial Ecology* 23: 77–95. [CrossRef]

Steen, Bengt. 1999. *A systematic Approach to Environmental Priority Strategies in Product Development (EPS): Version 2000-Models and Data of the Default Method*. Gothenburg: Chalmers University of Technology, Environmental Systems Analysis.

Suárez-Eiroa, Brais, Emilio Fernández, Gonzalo Méndez-Martínez, and David Soto-Oñate. 2019. Operational principles of circular economy for sustainable development: Linking theory and practice. *Journal of Cleaner Production* 214: 952–61. [CrossRef]

Sugiawan, Yogi, Robi Kurniawan, and Shunsuke Managi. 2019. Are carbon dioxide emission reductions compatible with sustainable well-being? *Applied Energy* 242: 1–11. [CrossRef]

Tăchiciu, Laurențiu. 2018. The Circular Economy between Desiderates and Realities. *Amfiteatru Economic* 20: 245–46. [CrossRef]

Tseng, Ming-Lang, Raymond R. Tan, Anthony S. F. Chiu, Chen-Fu Chien, and Tsai Chi Kuo. 2018. Circular economy meets industry 4.0: Can big data drive industrial symbiosis? *Resources, Conservation and Recycling* 131: 146–47. [CrossRef]

UNEP/SETAC. 2009. *Guidelines for Social Life Cycle Assessment of Products*. Life-Cycle Initiative. Paris: United Nations Environment Programme and Society for Environmental Toxicology and Chemistry, Available online: http://www.unep.fr/shared/publications/pdf/dtix1164xpa-guidelines_slca.pdf (accessed on 27 June 2019).

Urbinati, Andrea, Davide Chiaroni, and Giovanni Toletti. 2019. Managing the Introduction of Circular Products: Evidence from the Beverage Industry. *Sustainability* 11: 3650. [CrossRef]

Wernet, Gregor, Christian Bauer, Bernhard Steubing, Jürgen Reinhard, Emilia Moreno-Ruiz, and Bo Weidema. 2016. The ecoinvent database version 3 (part I): Overview and methodology. *The International Journal of Life Cycle Assessment* 21: 1218–30. [CrossRef]

Yeo, Zhiquan, Donato Masi, Jonathan Sze Choong Low, Yen Ting Ng, Puay Siew Tan, and Stuart Barnes. 2019. Tools for promoting industrial symbiosis: A systematic review. *Journal of Industrial Ecology*, 1–22. [CrossRef]

Zaman, Atiq. 2017. A Strategic Framework for Working toward Zero Waste Societies Based on Perceptions Surveys. *Recycling* 2: 1. [CrossRef]

Zucchella, Antonella, and Pietro Previtali. 2019. Circular business models for sustainable development: A "waste is food" restorative ecosystem. *Business Strategy and the Environment* 28: 274–85. [CrossRef]

© 2019 by the authors. Licensee MDPI, Basel, Switzerland. This article is an open access article distributed under the terms and conditions of the Creative Commons Attribution (CC BY) license (http://creativecommons.org/licenses/by/4.0/).

Article

Decomposing the Complexity of Value: Integration of Digital Transformation of Education with Circular Economy Transition

Serdar Türkeli [1],* and Martine Schophuizen [2,3]

[1] United Nations University–Maastricht Economic and Social Research Institute on Innovation and Technology (UNU-MERIT), Maastricht University, Boschstraat 24, 6211AX Maastricht, The Netherlands
[2] Welten Institute, Open University, Valkenburgerweg 177, 6419 AT Heerlen, The Netherlands
[3] LDE Centre for Education and Learning, Faculty of Electrical Engineering, Mathematics and Computer Science, TU Delft, Mekelweg 5, 2628 CD Delft, The Netherlands
* Correspondence: turkeli@merit.unu.edu

Received: 29 June 2019; Accepted: 15 August 2019; Published: 20 August 2019

Abstract: In this article, we highlight the pressing need for integrating the windows of opportunities that digital transformation of education opens up with circular economy education to accelerate the achievements of sustainability outcomes. Circular economy transition, as a multi-scalar process, relates to several contexts, e.g., product, firm, industry-level transformations ranging from designing local socio-technical solutions to greening global value chains, with multi-level policy and business implications for finance, production, distribution, consumption that are fundamentally consequential to everyday life, work and learning. Drawing on theories of neo-capital, multi-level perspective and structuration, and as methodology, using content analysis and qualitative meta-synthesis of scientific publications in digital education for sustainability, we blended our findings into multi-level, multi-domain structuration blueprints, which capture the complexity of value emanating from the interactions among external structures, internal structures of agents, active agencies and outcomes, for circular economy open online education and massive open online course instructional designs. We conclude that learning and creating multiple values to increase social–ecological value, complementarily to economic value, necessitate activating the complexity of value embedded in digital education and circular economy transitions with customizable niches of learning preferences and journeys of individuals and groups, within broader (and evolving) technological, organizational and institutional structures.

Keywords: sustainability; circular economy; education; digital transformation of education; open online education; open educational resources; massive open online courses; MOOCs; complexity; multiple value

---

1. Introduction

Transition to a circular economy (CE) requires a systemic change in an educational context and content at all levels and stages of teaching and learning (e.g., curriculum design, professional development, lifelong learning). The most pressing and complex issues of globally connected financing, production, distribution and consumption patterns of our times challenge policymakers, financiers, managers, administrators, educators and learners, as well as entrepreneurs, to ideate about and demonstrate various reliable regenerative solutions in various domains and disciplines (e.g., in sciences, design technologies, information and communication technologies (ICT), business models, finance, economics, environmental sciences, geography; Ellen MacArthur Foundation (2019)) in order to help accelerate linear systems to move beyond the destructive "take–make–waste" model in creative ways by creating multiple value propositions (for an extended discussion on value propositions, please see Corvellec and Hultman (2014)).

CE transition in this comprehensive scope also relates itself to several sustainable development goals (SDGs), targets and outcomes regarding SDG 6 Clean Water and Sanitation, SDG 7 Affordable and Clean Energy, SDG 8 Decent Work and Economic Growth, SDG 9 Industry, Innovation, Infrastructure, SDG 11 Sustainable Cities and Communities, SDG 12 Sustainable Consumption and Production, SDG 13 Climate Action, SDG 14 Life Below Water, SDG 15 Life on Land (United Nations 2018). However, in this article, we also emphasize the role of education, SDG 4 on Quality Education, especially, the importance of integrating the digital transformation of education with CE education and transition for achieving economic, social and environmental sustainability outcomes through CE open online education (OOE) and CE Massive Open Online Course (MOOC) instructional designs.

It is noted by many scholars that moving towards a CE consists of highly complex tasks and is an integrated action and process among ecological, social, economic, and even political dynamics (Bocken et al. 2017; Blomsma and Brennan 2017; Geissdoerfer et al. 2017; Geng et al. 2012; McDowall et al. 2017; Türkeli et al. 2018). Yet educational processes that are characterized by teacher-centered unidirectional instruction of the current socio-technical regime of education is not a competent fit for teaching and learning about CE, let alone developing tangible sustainable production and consumption solutions that are in line with a CE. The complexity of CE transition which introduces three knowledge constraints about circular economic systems, such as input-stage constraints (e.g., unknowns about the initial state of the linear and circular systems), process-stage constraints (e.g., unknowns about the process between input and output stages which also consists of unknown barriers to a CE system), and output-stage constraints (e.g., the targeted/desired state of an emerging CE system), makes pure theoretical teaching incompetent to capture the complexity of designing and implementing multiple value solutions for a systemic circular economic change while moving away from a linear economy.

While in CE literature, the use and the role of digital technologies are discussed in terms of new circular economic (digital) business models and technologies, CE digital education is overlooked (Türkeli et al. 2018). Moreover, although digital, online technologies, utilization of technology-enhanced learning (TEL) solutions, and the use of emerging digital technologies have started to demonstrate their potentials in transforming education (e.g., systems, modes, methods, activities relating to teaching and learning) (Papadakis 2016). The facilitation of these digital educational technologies for teaching and learning CE, and thus, supporting the acceleration of a transition towards CE, in practice, have also been and remain limited. Their potential has not been used to a large extent, even for education for sustainability, in an integrated manner (Zhan et al. 2015). Although micro-level adoption and application of educational technologies to make traditional teaching practices easier with a focus on utilizing more efficient ways to traditional teaching is an appropriate starting point, education for a CE transition remains underdeveloped. Currently, with a rather limited and closed—instead of open—use of available digital, online educational technologies, technologically advanced and enhanced ways of learning and teaching, potential benefits of digital transformation of education by CE open educational resources (OER), CE MOOCs and CE OOE remain significantly underutilized. In this sense, compared to unidirectional teaching, non-unidirectional teaching, interactive instruction (e.g., problem, peer, project, consultancy-based learning) and OOE initiatives that incorporate interactive instruction in a massive setting (e.g., connectivist massive open online courses (cMOOCs) concentrating on knowledge co-construction, contrary to xMOOCs) are more promising in being able to create multiple value solutions and capture the dynamism of interactive learning processes, which are inherent to the interdisciplinary characteristics of real-world sustainability issues and solutions, thus, education (Wals and Jickling 2002; Steiner and Posch 2006). Yet, there are no studies in particular, focusing on the educational and instructional design needs of such OER, OOE and MOOCs for CE teaching and learning.

Supported by the advance of information and communication technologies (ICTs), as well as networking technologies; e-learning, OOE, TEL, OER and MOOCs operate on virtual learning environments (VLEs) and networks benefitting from physical capital such as the availability of technological infrastructures, networking and incumbent technologies in use and emerging technologies in development. While digital technologies enable economic growth as they are being applied within a wide range of sectors and transform many organizations at an increasingly fast pace (Orlikowski 2000), digital transformation of education globally is still a socio-technical niche. Moreover, further educational informatization, networking and emerging education technologies day by day become inevitable (e.g., networked OERs, MOOCs, and VLEs). Digital technologies, as a source and/or enabler of innovation, continue changing and cumulatively transforming the form and functioning of the organization of education, educational organization, and thus, education itself. The emerging agreement or consensus on the positive effects of digitization of (and within) education actually benefits from further enabling and rapid transfer of (big) data in various formats (text, image, video, sound), information and (codified) knowledge. Also, emerging digital technologies have the potential to make education more flexible, as they allow for learning anywhere and anytime (e.g., through forms of OOE, seamless learning). Indirectly, they, therefore, also help to increase the possibility of access to education. These, among other technological possibilities, accelerate the use of technology and especially, newly emerging ICTs in education. In its basic sense, CE OOE can present several starting attributes of the new pedagogical models via integrating gamification, as well as high-impact short videos, infographics and OER in the presentation of CE content. These novelties are supporting elements while integrating a multidisciplinary team around several CE-related issues and tasks. Activities are then defined by collaboration, communication and commitment of the educators and learners and enabled, monitored and/or evaluated processes by emerging technologies. While all these socio-technical developments go well together with the idea of a knowledge society, including an attention to lifelong learning in a context, they also demand continual updating of human knowledge and skills, as well as continuous research and development of new educational technologies. The process of designing and implementing such open, collaborative and multidisciplinary infrastructures and technologies are prone to many challenges, yet carry progressive potentials (Schophuizen et al. 2018).

Considering the systems thinking at its core and the complexity of CE transition as a systemic transformation, among others, the theory of online collaborative learning (OCL) provides us a correspondingly multifaceted basis for operationalizing complex teaching and learning in transformative ways (Scardamalia and Bereiter 2006). OCL, unlike the classical behaviorist and cognitivist focus on instructions for replication of facts, emphasizes the process of building knowledge (Scardamalia and Bereiter 2006). In this respect, it also differs from constructivist learning by locating active learning within a process of social and conceptual development based on knowledge discourse (Harasim 2017). Online and technology-enhanced interactions then enable an active learning space via two major actors, the learner and the teacher, and their (technology-enhanced) interactions in learning, and with content matter (Anderson 2004). This process and interaction-based approach fits with teaching and learning CE and to build CE solutions since a community of inquiry, with synchronous and asynchronous activities (e.g., video lectures, podcasts, conferencing, chats, or virtual global interactions), if designed well, constitutes several technology-enhanced online learning environments which support rich interactions for learning and building novel knowledge content, and developing social skills through development of personal relationships among participants, which eventually lead to online collaborative co-construction of knowledge within a community of inquiry (Gunawardena et al. 2006). For education in ill-structured knowledge domains (e.g., education for and on a CE), an instructional design in which active learning and knowledge building can be combined, and a community of inquiry in which learner support, mentoring and knowledge innovation can be provided, are well suited due to the fact that this setting allows for perspective transformations as an end goal, which occurs (and required) at both the individual and community.

Yet, the question remains which specific modes, methods and activities are most used and found suitable for teaching and learning for a CE transition. As we indicated above, while the use and the role of digital technologies in education have received increasing attention from scholars, and an emerging attention for education for sustainability, they have still yet to emerge for CE education (Türkeli et al. 2018). Therefore, the current challenge we address in this article is to investigate and discuss the ways (modes, methods and activities), in short, the constituents, which are, in principle, deemed and discussed to be most effective, in creating and distributing quality TEL solutions for accelerating the transition towards a CE.

Considering the aforementioned complexities of both research fields, the main research question of this article is: What are the ways in which the constituents of structuration dynamics among various agents, technologies and structures contribute to the extents to which the digital transformation of education has been utilized in education for sustainability, and how they relate to and can be further utilized for circular economy education to accelerate a transition towards a circular economy?

To answer our research question and to position our contribution in the current state of the research fields and literature in CE and OOE, the empirical evidence needs to be processed and appraised with respect to several theoretical and conceptual lenses. While decomposing the complexity of value emanating from integration of digital education with CE transition, we draw on theories of neo-capital, multi-level perspective and structuration theory (Section 2.1). As methodology, we use content analysis and qualitative meta-synthesis of scientific publications in the field of digital education for sustainability, in particular, the scientific publications (n = 36) concentrating on OOE and MOOCs in the field of education for sustainability (for details on data source, data and methodology, please refer to Sections 2.2 and 2.3). We present our findings in Section 3, which are further discussed in Section 4. We provide our conclusion, and future research directions in Section 5.

2. Materials and Methods

2.1. Materials: Theoretical Lenses

We draw upon three theoretical lenses (theories of neo-capital (Section 2.1.1), multi-level perspective (Section 2.1.2) and structuration theory (Section 2.1.3)) to be able to decompose the complexity of value (theories of neo-capital) into multi-level (multi-level perspective) and multi-domain (structuration theory) into interrelating analytical categories for the integration of digital transformation of education with CE education, and thus a CE transition. By taking various varieties of capital (value) formation at individual and group level, varieties of agents, emerging as well as broader and evolving technological, organizational and institutional structures into account, we present our multi-level and multi-domain findings in tables, discuss them, and build a multi-level and multi-domain thematic synthesis for CE OOE and CE MOOC designs. In the following subsections, each theoretical lens with their conceptual apparatus is positioned and substantiated with respect to our research question, research goal, presentation of the results, and follow-up discussions, conclusions and future research directions around CE OOE and CE MOOC instructional designs.

2.1.1. Neo-Capital Theories: The Necessity of Varieties of Capital Formation by Digital Education for CE Teaching and Learning

Due to their inherent multi-dimensional complexities, digital transformation of education (Anderson 2004; Scardamalia and Bereiter 2006; Gunawardena et al. 2006; Cano 2015; Harasim 2017; Schophuizen et al. 2018) and CE transition (Geng et al. 2012; Bocken et al. 2017; Blomsma and Brennan 2017; Geissdoerfer et al. 2017; McDowall et al. 2017; Türkeli et al. 2018), and especially, a synergetic integration of them is not a straightforward endeavor. Taking different material interests and discrete ideas of various agents in teaching and learning CE into account, this integration calls for and demands varieties of capital formation, and capital-intensive efforts from a multitude of agents (e.g., policymakers, administrators, educators, entrepreneurs, intrapreneurs), institutions and organizations (e.g., public, private and social sector organizations). The varieties of capital formation exceeds (co-)investing in financial and/or physical capital (e.g., funds, subsidies, infrastructures and technologies) and necessitates (co-)developing relevant human capital (e.g., the skills, knowledge), social capital (e.g., access and use of connected/networked resources such as data, information, tools, courses), cultural capital (e.g., norms, mindset, attitudinal trainings) in designing, improving and implementing novel CE VLEs with a view on novel natural capital formation (e.g., novel waste-to-resource and energy transformations for bio-ingredients and technical materials) (Lin 2017; Stahel 2019; Caruso and Gattone 2019; D'Adamo 2018). Systems thinking being at its creative core, such circular economic system-wide interactions among agents and structures necessitate creating and sustaining aforementioned varieties of (neo-)capital formations and economic, societal, environmental returns via extending towards experiential learning activities (e.g., see Kolb 2014), interdisciplinary approaches and tasks, multi- and transdisciplinary inquires, projects, collaborations, demonstrations and commercialization activities for a CE transition. Classical, traditional teaching, learning philosophies, modes, methods and activities, in short, the constituents of the current socio-technical regime of education, are challenged daily, even in terms of being able to deliver reliable regenerative sustainability solutions for an integrated economic, societal, and ecological survival, let alone, welfare (UNESCO 2015). Thus, in this article, as levels to varieties of capital formation, we concentrate on socio-technical transitions and transition management literature (Geels 2004; Geels and Schot 2007; Kemp 2011; Geels 2011), in particular, on the multi-level perspective (MLP) which assesses socio-technical transitions and change by focusing on the interrelationship of three analytical levels:

(1) Socio-technical niche innovations at micro level (e.g., related agents (educators, learners, entrepreneurs ...); technologies (CE OER, CE MOOCs, emerging education technologies ...); structures (e.g., classroom, course, university ...)),
(2) Socio-technical regimes at meso level (e.g., related agents (policy makers, administrators, businesses ...); technologies (e.g., CE OOE, incumbent education technologies ...); structures (e.g., higher education system, public education policies, labour market conditions ...)),
(3) Socio-technical landscapes at macro level (e.g., related agents (politicians, policymakers, businesses ...); technologies (e.g., next-generation OOE technologies, technology visions ...); structures (e.g., international organizations, OOE and/or CE policy visions ...)).

The background and relevance of MLP are provided in the following subsection.

2.1.2. Multi-Level Perspective: Niche, Regime and Landscape-Level Transitions for Digital Education for CE Teaching and Learning

MLP, as an analytical tool, conceptualizes overall dynamic patterns in a socio-technical transition, the complexity of change, and resistance to change at the socio-technical regime level (Geels 2011). Socio-technical transitions in this sense are the fundamental shifts in the socio-technical systems (Geels 2004; Geels and Schot 2007; Kemp 2011). Such transitions necessitate changes in material, technological, organizational, institutional, political, economic and socio-cultural dimensions (Markard et al. 2012). Change is both co-initiated from bottom-up via socio-technical niche innovations (micro, e.g., CE OOE, CE MOOCs) (which form the micro-level, are incubated from regulatory and market pressures, protected from the rules of the dominant socio-technical regime regarding varieties of capital formation and continuity, or not, or not yet) and socio-technical landscape developments (macro, e.g., OOE policies and visions, see European Commission (2013), CE policies and visions, and McDowall et al. (2017)) which exert pressure on socio-technical regimes (meso). A socio-technical regime in this sense is an institutionalized set of rules that organizes the actions of agent groups who, in turn, reproduce or transform the constituents of a socio-technical system (Geels 2004, 2011). The macro level, the socio-technical landscape, is the broader environment that influences both the socio-technical regime and niche innovations. These landscape factors are beyond the direct control of agents (Geels 2004, 2011; Geels and Schot 2007). In this regard, CE education, digital transformation of education, digital education for CE, and CE transition are all socio-technical niche areas in which the integration of digital transformation of education with teaching and learning for a CE transition would challenge the status quo—the socio-technical regime of education. Particularly, what the current socio-technical regime of the education system, as a socio-technical system, offers for sustainability, and in particular, CE education, are institutionalized set of rules and structures that influence the actions of agents and agent groups, who in turn, reproduce or transform the constituents of the current education socio-technical system. Thus, we utilize structuration theory to reveal such multi-level and multi-domain structuration dynamics over and on the constituents of change. We elaborate on our findings (Section 3) in detail in Section 4. The next subsection provides the quadripartite nature of structuration throughout mutual shaping processes between agency and structure, which is utilized in framing and revealing the multi-level and multi-domain structuration dynamics over and on the constituents of change towards CE OOE.

2.1.3. Quadripartite Nature of Structuration: Mutual Shaping of Agency and Structure through the Constituents of Change towards CE OOE

Giddens (1984) proposed the concept of structuration to capture and explain the mutual shaping and interactions between agents and structures. In his constructive critique, Stones (2005) pointed out the quadripartite nature of structuration (Figure 1):

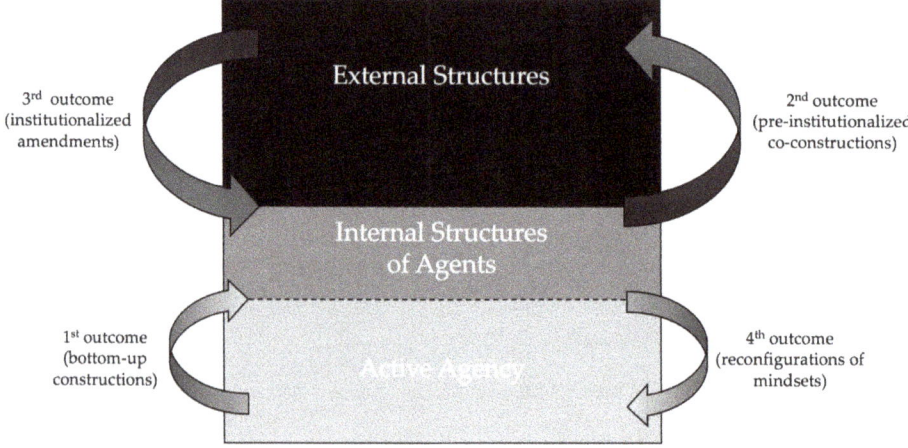

Figure 1. Quadripartite nature of structuration. Source: Authors' work.

External structures of structure (autonomous from the agents in focus): External structures form the conditions of actions for agents. They are top-down or else hierarchical external structures that come with institutionalized set of rules, norms and values, reflected in (e.g., OOE and/or CE) policies, programmes, plans, politico-administrative, organizational strategies, set-in rules and allocated resources. Structures have a dominant institutionalized design encouraging and/or challenging the continuity and/or coherence of actions of agents via institutionalized amendments (towards transformation) or institutional repudiation (reproduction) (Figure 1).

Internal structures of agents: Field corresponds to conjuncturally specific knowledge of external structures within position–practice relations (e.g., action-informing assessment of relational norms, power relations, action-informing conclusions of relevant (and/or networked) agents-in-context) and general-dispositions (Stones 2005), or following Bourdieu (1972), habitus, (e.g., generalized world-views, frames, cultural schemas, classifications, associative chains and connotations of discourse methodologies for adapting generalized knowledge to a range of particular practices in particular locations in space and time). These internal structures (internal structures within each agent and community of agents) are based on the evidence of positively or negatively institutionalized perceptions and perspectives (normative and cognitive rules and resources) of professional agents (e.g., policymakers, administrators, entrepreneurs, educators, learners about OOE and/or CE), of professional bodies (e.g., legislative, executive, accreditation, standardization bodies) and/or networks of such agents and bodies (Figure 1).

Active agency: Active agency includes the ways the agents, either routinely, pre-reflectively or strategically and/or critically draw upon their aforementioned internal structures, and take action (Stones 2005) (Figure 1).

Outcomes: Outcomes are events through which external structures and internal structures of agents evolve (Stones 2005). Outcomes are bottom-up, constructed, pre-institutionalized amendments, proposals for gradual change in external and internal structures, if institutionalized and aggregated, they are the constituents of an accumulated change, top-down (e.g., novel understanding/mindsets, policies and visions, strategies for creating incentives which include novel rationales, rules and resources for OOE and CE) (Figure 1).

2.2. Materials: Data Source and Data

Three main data sources for detecting peer-reviewed, high-quality scholarly publications of scientific and international relevance are Web of Science (WoS) of Thomson Reuters, Scopus of Elsevier, and Google Scholar of Google Inc. (Türkeli et al. 2018). We chose using the WoS scientific citation indexing service due to the facts that WoS has the most strict quality criteria for selection and inclusion of scientific publications in its indexing services as a trusted and authoritative source of bibliometric data and information for peer-reviewed global research knowledge across disciplines, for a comparison of these three data sources, please refer to Falagas et al. (2008), Harzing and Alakangas (2016) and Tennant et al. (2019). WoS provides a curated collection of over 21,177 peer-reviewed, high-quality scholarly journals and over 74 million records published worldwide, including open access journals in over 250 science, social sciences, and humanities disciplines (Tennant et al. 2019). To be able to capture the broadest range of peer-reviewed scientific activity and all international publications of scientific relevance in the field of education for sustainability, our unit of analysis is a scientific publication, which can be an article, a proceedings paper, book review, editorial material, letter, meeting abstract, review, in all languages available, and English translations of abstracts. Our WoS Core Collection query with the search term "education for sustainability", which is an umbrella term for all forms of education that promote rethinking of educational programmes (contents and methods) and systems for sustainable societies, returned 386 international publications of scientific relevance. In order to detect the relevant subset for the digital transformation of education within these high-quality scientific publications around the world, this dataset is then refined by a combinatorial keyword query set such as: "digital education", "online", "online education", "open education", "open online education", "massive open online course", "MOOC". In total, we retrieved 36 scientific publications by 29 March 2019 with these strict quality and query criteria (please refer to the supplementary file for an extensive list of publications).

2.3. Method: Qualitative Metasynthesis

We applied qualitative meta-synthesis on these 36 scientific publications. Qualitative metasynthesis is a research approach to analyze data across scientific publications. The intention is to reach a coherent state of structured information which enables researchers to classify findings of earlier scientific studies, as latent indicators, as constituents and their building blocks, of a latent outcome, while interactions of which, manifest or do not manifest a hypothesized meta-model. In this respect, qualitative metasynthesis as an interpretive analytical technique relies on content analysis and uses the findings reported in previous scientific publications, as constituents and their building blocks for gaining a deeper understanding of particular phenomena, and to build up a new thematic synthesis (Finfgeld-Connett 2014).

We followed a five-step qualitative metasynthesis process (Erwin et al. 2011) to reach our research goal by providing answers to our research question. These steps are:

1. Framing (WoS Core Collection query: "education for sustainability") (Section 2.2),
2. Searching (Refinement of the WoS results (n = 386 scientific publications) with our keywords of interests "digital education", "online", "online education", "open education", "open online education", "massive open online courses", "MOOC", as the query is often refined and reduced in scope over the course of undertaking the qualitative meta-synthesis (please refer to Walsh and Downe 2005) (Section 2.2),
3. Selecting and appraising the relevant scientific publications (manual check of the relevance and reading of remaining 36 scientific publications with respect to the sample quality criteria (please refer to Atkins et al. 2008)),
4. Summarizing and synthesizing (gathering findings, constituents and their building blocks from resulting scientific publications either falling in or outside the theoretical lenses used in this article (Section 2.1)),
5. Combining and reporting the evidence by identifying the key themes/concepts in each study via lines-of-argument synthesis, while developing a general interpretation of the phenomena of interest (e.g., CE OOE, CE MOOC instructional designs) that is grounded in the themes/concepts of each study. As a result, we generate analytical themes that emerge from, and step beyond the descriptive themes as a new thematic synthesis (Thomas and Harden 2008). This last step constitutes the building blocks (Tables in Section 3) for our research, and addresses our research question.

The overall process synthesizes the constituents of existing findings of earlier scientific publications to construct a greater meaning through an interpretative process, discussions and concluding remarks which intellectualize an abstract phenomenon into a tangible proposal (Sections 3–5). In this regard, this article exceeds the category of a systemic literature review, and incorporates original research article elements in its endeavor, and allows other researchers to replicate and build on these published results.

3. Results

3.1. Structuration Constituents among Classical Capital and Neo-Capital

We start our analysis by presenting the findings relating to the structuration constituents among classical capital and neo-capital.

Our findings indicate that a close alignment of the course topics and subject matters with learners' personal ideas and material interests, by supplying and/or offering them a set of attractive value propositions regarding the demanded time and financial commitments compared to formal education or training (Grealy 2015; Leire et al. 2016; Howarth et al. 2016; Meinert et al. 2018) is among the most important determinants for a competent OOE and/or MOOC instructional design in digital education for sustainability. However, exceeding the time and financial commitments, our findings (n: 22) justify that these value propositions extend to various structuration dynamics over and on the constituents relating to both classical capital and neo-capital (Table 1). Our findings cluster around classical capital needs (e.g., money/time, physical, technological capital (five findings)) and neo-capital formation supply by educators and demand needs by learners (human capital (seven findings), social capital (six findings), and cultural capital (four findings)). We further discuss these results in Section 4.

Table 1. Structuration constituents among classical capital and neo-capital.

Varieties of Capital Needs for CE OOE and/or CE MOOCs	Constituents	Source in EfS
Human Capital Development Focus: Internal structures of individual agents and active agency of individual agents (seven findings)	Autonomous study and the sharing of global education resources	(Li and Zhou 2018)
	Autonomy in, mastery of and purpose of using the tools (e.g., skill-building) via MOOCs	(Fini 2009)
	Possibility of developing personal knowledge management skills with options for passive, time-saving mailing lists and interactive, time-consuming discussions forums	(Fini 2009)
	Setting up a right balance between theoretical and practical examples by ensuring satisfaction with case studies	(Aksela et al. 2016)
	Changing and improving personal perceptions of sustainability	(Aksela et al. 2016)
	Improving the understanding of complex systems (e.g., interrelatedness of CE transitions and Sustainable Development Goals)	(Aksela et al. 2016)
	Providing other support (e.g., technical and learning strategies for learners)	(Aksela et al. 2016; Cano 2015)
Social Capital Development Focus: Internal structures of groups of agents and active agency of groups of agents (six findings)	The support for interaction, the integration of a multidisciplinary team around an issue	(Griselda Argueta-Velazquez and Ramirez-Montoya 2017; Ramirez-Montoya 2018)
	Communication with other learners and getting feedback from them, as well as from teachers and tutors	(Aksela et al. 2016; Mishra et al. 2017)
	Possibility of establishing interdisciplinary MOOC study groups	(Chen and Chen 2015)
	Making supporting tools available to convene a conversation about, e.g., circular economy, sustainable development, sustainability, and climate change	(Burch and Harris 2014)
	Maintaining an active participant discussion even in the case of the removal of educator facilitation	(Sneddon et al. 2018; Aksela et al. 2016)
	Connectivist–heutagogical (social and cultural individual) learning using Garrison's Community of Inquiry Model, e.g., (cognitive, social and teaching presence for educational experiences)	(Kaul et al. 2018)
Cultural Capital Development Focus: Internal structures of agents and active agency of agents and groups of agents (four findings)	Supporting and logging the processes of negotiation, cultural articulation, and identity formation which occur through e-conversations and which can include large populations from different backgrounds	(Burch and Harris 2014)
	Activating the implications of these e-conversations for the broader, e.g., climate change, discourse for the definition of the problem, attributions of responsibilities, and the development of solution offers, options or solutions	(Burch and Harris 2014)
	Improving the understanding of the complex systems (e.g., interrelatedness of, e.g., CE transitions and SDGs) in communities	(Aksela et al. 2016)
	Changing and improving perceptions of sustainability in communities	(Aksela et al. 2016)

Table 1. *Cont.*

Varieties of Capital Needs for CE OOE and/or CE MOOCs	Constituents	Source in EfS
Physical/Technological Capital Development **Focus: External Structures** (Organizational, Technological, Institutional) (five findings)	Presentation of educational content through high-impact short videos, info-graphics and OER	(Griselda Argueta-Velazquez and Ramirez-Montoya 2017; Ramirez-Montoya 2018)
	Using a pedagogical model that integrates gamification into teaching and learning	(Griselda Argueta-Velazquez and Ramirez-Montoya 2017; Ramirez-Montoya 2018)
	Refining the socio-technical issues of computer-supported collaborative learning	(Wise and Schwarz 2017)
	Providing mobile MOOCs, the mobile-based programs, apps, which enable teachers to use the program without space and time constraints or providing learners on the move a seamless accessibility to content, both of which are deemed as key enabling factors in post-qualification and continuing professional development courses for busy practitioners who work full-time.	(Meinert et al. 2018; Grealy 2015)
	Technological tailoring for particular learning needs of solicitors and trainees who have time-demanding careers, and who would benefit from being offered flexible options in terms of engaging with their learning processes, supported by digital notifications	(Meinert et al. 2018)

Source: Authors' compilation, EfS: Education for Sustainability.

3.2. Structuration Constituents among Incumbent Technologies and Emerging Technologies

Following the synthesized presentation of structuration dynamics over and on classical capital and neo-capital formation needs for CE OOE and/or CE MOOC instructional designs (Table 1), we present the results for the structuration constituents among incumbent educational technologies and emerging technologies. We have identified 11 structuration constituents among emerging technologies and incumbent technologies that are focusing on various agents (learners, educators, managers, entrepreneurs, technologies and their interactions) (Table 2). While learning analytics using neural networks, G-Rubric, semantic analysis models focus on learners, SMART Teaching 3.0 and Creative Commons (CC) licenses for teaching material development focus on educators. These findings indicate the relevance of emerging technologies for both educators and learners for a CE OOE context and CE MOOC designs. Moreover, the focus also extends to improving technologies themselves and the interactions between technologies and human agents (e.g., Fedora Commons Repository, FedX API, Moodle and Elgg integration).

Table 2. Structuration constituents among incumbent technologies and emerging technologies.

Emerging Technologies	Agents in Focus	Constituents	Source in EfS
Learning Analytics using Neural Networks	Learners	Static versus Dynamic: A hybrid neural network (NN) model which integrates a Convolutional Neural Networks (CNN) and with Gated Recurrent Unit (GRU)-based Recurrent Neural Networks (RNN) in an effort to dynamically detect individual learning features	(Li and Zhou 2018)
G-Rubric	Learners	Assessment tech versus semantic assessment technologies, via using latent semantic analysis (LSA) as an automatic assessment tool	(Santamaria Lancho et al. 2018)
Semantic Analysis Models	Learners	Transition from impassive analysis models to a semantic analysis model (SMA) to track the emotional tendencies of learners	(Wang et al. 2018)

Table 2. *Cont.*

Emerging Technologies	Agents in Focus	Constituents	Source in EfS
SMART Teaching 3.0	Educators	Sharing educational experiences of in-service teachers using the community of inquiry (CoI) framework, providing community-centered professional support for in-service teachers	(Jin-Hwa and Kim 2016; Kaul et al. 2018)
Creative Commons Licenses (CC)	Educators (Licensing)	A paradigm shift from top-down, institution-centered teacher training to bottom-up, learner-centered professional development in teacher education with reuse and redistribution among instructors due to the privacy of the contents created. Emerging issue: intellectual property right protection of networked teaching and open educational resources	(Jin-Hwa and Kim 2016)
Fedora Commons Repository	Technology (Managers, entrepreneurs)	Repository for an OER back-end to manage OER resources	(Chunwijitra et al. 2016)
Moodle and Elgg	Technology Managers, Entrepreneurs (Interactions)	MOOCs which imply an integration of virtual learning environments	(Coelho et al. 2015)
FedX API	Technology Managers, Entrepreneurs (Interactions)	An API including a packet encapsulation and a data transmission module which organizes open educational resources between systems. Resources can be exchanged among the third-party OER repositories by an OAI-PMH harvesting tool, situating an OER-MOOC interaction	(Chunwijitra et al. 2016)
OER-MOOC Interaction	Learners, Technology, Educators, Entrepreneurs (Interactions)	Development of online educational resource sharing and analyzing these sharing activities through a comparative analysis of foreign open classes and domestic resource sharing courses, sharing of resources and communication between teachers and students; Open Educational Resources, Open Courses, Open Communities and Open Schooling	(Cai et al. 2016; Okada and Sherborne 2018)
WeChat Public Platform Integration	Learners, Technology, Educators (Interactions)	Improving learning efficiency via positive interactions between teachers and students in a network-based teaching mode using the WeChat public platform to build a virtual learning environment that comes with standards for public educational resources	(Cai et al. 2016)
Digital Learning Strategies	Learners, Technology, Educators, Entrepreneurs (Interactions)	Situating technology-enhanced learning strategies: such as content curators, information filters (proposing systems), learning algorithms for intelligent and self-adaptive tutorial systems as novel digital learning strategies	(Cano 2015)

Source: Authors' compilation, EfS: Education for Sustainability.

These emerging interactions and connectedness features for learners, educators, managers, entrepreneurs and technologies (e.g., OER-MOOCs interactions, public platform integrations, situating digital learning strategies via TEL strategies) are relevant for a CE OOE context and CE MOOCs instructional design due to the multiplicity of domains (e.g., engineering, design, economics, business perspectives) and systems thinking involved in CE (Ellen MacArthur Foundation 2018). We discuss these results in Section 4.

3.3. Structuration Constituents between Agents and Structures

Following the structuration constituents among traditional, incumbent technologies and novelties that emerging technologies promise (Table 2), we present multi-level and multi-domain structuration dynamics between (varieties of) agents and (varieties of) structures (Tables 3 and 4). Our findings justify the existence and the importance of varieties of structures (e.g., structure of technological platforms, learning management systems, technological areas, course structure, evaluation structure, societal structure), internal structures of agents (e.g., relevance for agents, learning preferences, motivations of learners), active agency situations (e.g., autonomous study, embedding innovations in education, active involvement), and outcomes (e.g., awards for participation, sharing resources, developing digital scientific literacy). Due to the inflation in the number of potential policy, business, environmental and social issue niches, interventions and innovation entry points for a CE, niche socio-technical CE solutions require such contextualized, personalized, customizable, flexible and self-directed ways of learning with authentic learning tasks and assignments (Table 3). In a CE OOE context and for CE MOOCs instructional designs, these dynamics between actors and structures should be taken into account to activate agency, and improvements, thus, the evolution of broader structures involved in CE OOE and CE MOOC instructional designs (Table 4). We provide a synthesis of these micro-level, meso-level (Table 3) and, broader and evolving macro-level and macro domain constituents in Table 4 with respect to various agents and structures involved. We discuss these results in Section 4.

Table 3. Structuration constituents among agents and structures.

External Structures	Internal Structure of Agents	Active Agency	Outcomes	Source in EfS
Major MOOC platforms, Learning Management Systems (Technological structures)	Individual learning preferences	Autonomous study (Learners)	Sharing of global education resources, the MOOC platforms offering specific learning paths, relevant contents individually according to learners' identified learning features	(Li and Zhou 2018)
Societal structure relevance	Vocational relevance	Individual relevance (Learners)	Autonomy, Competence/Mastery, Purpose	(Aksela et al. 2016)
Course structure	Motivation by peers which increases continuity of interactions	Individual participation (Learners)	Improvement of online learning group processes	(Mayende et al. 2017; Gil et al. 2018)
Mandatory marks (Evaluation structure)	Virtual handholding	Active involvement (Learners)	Awards for participation	(Grealy 2015)
The areas of technology	Academic cultural practices	Embedding innovations (Educators)	Digital scientific literacy	(Fitzgerald et al. 2015)

Source: Author's Compilation, EfS: Education for Sustainability.

3.4. Structuration Constituents among Broader Set of Agents and Structures with a View on Technologies

At the global socio-technical landscape development level, the United Nations Decade of Education for Sustainable Development (DESD) calls for a more coherent, critical and multi-level analysis of learning environments in relation to creating seven competences (e.g., embracing diversity and interdisciplinarity, foresighted thinking, interpersonal competence, normative competence and systems thinking competence) (Mochizuki and Fadeeva 2010).

Mutual shaping of competences among each individual and groups, and such learning environment systems (Mochizuki and Fadeeva 2010), and also mutual shaping of competencies among each individual and groups, and global citizenship and sustainable development education based on competencies (Adomssent et al. 2007; Mannion et al. 2011) open up an investigation, analysis and discussion arena with an initial need of revealing and presenting further structuration constituents and dynamics among a broader set of (varieties of) agents and structures with a view on (varieties of) technologies (Giddens 1984; Orlikowski 2000; Stones 2005) (Table 4).

In this respect, the integration of digital transformation of education with a CE education, thus with an aim of accelerating CE transition, would systemically challenge the current socio-technical regime meso level, socio-technical systems of (CE) education, related organizations, policies, markets, technologies, and agents.

Table 4. Structuration dynamics among actors, structures and technologies.

Partite	Structuration Constituents with Respect to Technologies	Actors	Structures
External Structures of Structure (three themes)	(1) Politico-administrative public system, policy and programming (rules and resources) on CE, OOE, TEL, OER, SGs, MOOCs (2) Educational and technical administrative implementation plans and strategies (rules and resources) on CE, OOE, TEL, OER, SGs, MOOCs (3) Dominant curricular, instructional, and technological design structures and trajectories of CE, OOE, TEL, OER, SGs, MOOCs	Policymakers, Ministerial bureaucrats, CE and OOE experts Educational and Technical administrators, Educational technology developers, Educators, Learners, CE agents, Community organizers	Politico-administrative structure, Organizational technical structure, Techno-economic structure, Socio-economic structure
Internal Structures of Agents (two themes)	(1) Existence of and evidence for positively institutionalized perceptions (cognitive rules and resources) for CE, OOE, TEL, OER, SGs, MOOCs by professional agents (policymakers, ministerial bureaucrats, educational and technical administrators, educational technology developers, entrepreneurs, community organizers, teachers, learners); (2) Existence and evidence for accreditation, standardization bodies (and/or networks of such bodies and agents) relevant for OOE, TEL, OER, SGs, MOOCs in the context of CE education	*Additional to the list above:* Students, Learners, Teachers	*Additional to the list above:* Labor market structure
Active Agency (two themes)	(1) Existence of and evidence for acting institutional, educational and curriculum architects, CE and OOE entrepreneurs and intrapreneurs at the boundary of education and technology development systems who facilitate educational and technological change/transformation via or within OOE, TEL, OER, SGs, MOOCs in the context of CE (2) Existence of and evidence for involvement of users in OOE, TEL, OER, SGs, MOOCs; in teaching (teachers, instructors, educators) and in learning (students, learners, professionals) MOOCs in the context of CE	*Additional to the list above:* Entrepreneurial OOE and CE agents in public, private and social sectors	*Additional to the list above:* Labor market structure
Outcomes (two themes)	*Events* that relate to implementation of strategies (rules) for creating incentives (resources) and identification of best educational (technology) practices e.g., in: (1) Science, Technology, Engineering, Mathematics (STEM) subjects, (2) Broader disciplines (e.g., social sciences, design technology, ICT, business, finance, economics, environmental sciences, geography) for CE solutions by introducing OOE, TEL, OER, SGs, MOOCs	Agents and technologies with new rules, resources from organizations (of structures involved) New mindsets (for agents involved) New educational tools and technologies	Geographically, historically institutionalized chains/trajectories of survived/fittest outcomes, and events

Source: Authors' Compilation, SGs: Serious Games.

We provide a synthesis of these broader and evolving multi-level and multi-domain structuration constituents in Table 4 with respect to a broader set of agents and structures with a view on varieties of technologies. We discuss these results in Section 4.

4. Discussion

4.1. Multiple Value Formation through CE OOE and CE MOOC Instructional Designs

According to our findings, high rates of registration to online courses and high rates of completion require ownership from the learner side. Yet, we argue that in order to activate an ownership at the learners' side, multi-dimensional (individual and societal) value relevance in terms of varieties of neo-capital formation is needed. Necessities of human capital formation (e.g., relevant knowledge, skills) and social capital formation (e.g., interactions via informal learning, harnessing the collective intelligence of the learners, the interactions among other users such as former learners, future prospective learners, business professionals, universities, and organizations) should be taken into account and be provided to learners in a CE OOE context and CE MOOC instructional designs (Table 1). These enhancements help pave the way towards cultural capital formation (e.g., shift in mindsets, creating and adopting (new) ideas, establishing (new) interests). Elaboration on and implementation of customized teaching for learners from diverse backgrounds, different learning preferences, different language barriers, space and time constraints, ICT skills and at different stages of their learning (e.g., adolescents' learning, informal learners, lifelong learning adults) also require various different CE MOOC instructional designs that should come with various portfolios for varieties of neo-capital based multiple value formations possibilities and proposals such as a supporting form to and function of choice-based learning. In this respect, in Table 1, following our 22 main varieties of capital and neo-capital structuration constituents, we argue that, due to several issue niches that are spread throughout the multi-scalar technical material loops and bio-material loops of a CE in different sectors at different levels of a CE system, CE OOE and/or designing CE MOOCs as socio-technical innovation niches require proposing multiple value formation possibilities to many different issue agents (learners). Following our findings in Table 1, it is important to re-emphasize that educational enrolment through the use of technological tools and access to OERs should ensure that customizable varieties and portfolios of neo-capital based value formation possibilities are supplied by the educators, to meet the requirements of the learners. If pedagogical and evaluative treatment of the courses with acceptable educational parameters (e.g., quality versus massiveness) and the components of innovative attributes (e.g., integration of OER and gamification) are also in the educational and instructional designs of CE OOE and/or CE MOOC, these can further support the formation of customized learning portfolios which, in turn, by tapping onto the renewed interests and ideas of learners as agents, can help further shape the structural expansions of the organization of resources, tools, courses, and technologies involved in teaching and learning CE.

4.2. Emerging Educational Tools and Technologies for CE OOE and CE MOOC Instructional Designs

CE OOE and CE MOOC instructional designs require interdisciplinarity, online and offline networking, co-construction of knowledge between and among educators and learners. This type of co-construction is substantiated among teachers to increase digital literacy of teachers, targeting both pre-service teachers and in-service teachers and sharing educational experiences of in-service teachers in Oyo et al. (2017) or with community-centered professional support for in-service teachers in Kaul et al. (2018) (Table 2). We argue that through co-construction of innovations (e.g., via interacting OER, MOOCs, VLEs), of processes (e.g., interacting trajectories of teaching and learning) and of outputs (e.g., institutionalized VLEs, development of digital abilities and skills in addition to the learning (Carrera and Ramírez-Hernández 2018)), co-construction of sustainable, circular products, processes, organizations, and even a prospective economy, only then become truly targeted and possible with a contribution from CE OOE substantiated by several various CE MOOC instructional designs. At micro level, by putting the main agent—the learner—in focus, new analysis models, such as semantic analysis models (SMA) to track the emotional tendencies of learners become highly relevant (Wang et al. 2018). Additionally, since the learner is central, looking for the ways in how we facilitate the learning process in the teaching that is provided (e.g., supporting self-regulated learning by dashboards or widgets) also become central. However, not only the learner has a responsibility to take, but also, the teacher needs to explicitly consider these learner traits in designing education and should be aware of the targeted learner population and their specific needs. In this respect, we argue that active learning space of OCL theory becomes relevant for CE education. As indicated in Table 2, digital transformation of education, especially technology-enhanced educational designs for CE OOE via CE OER and/or CE MOOC instructional designs, and the use, development, and interactions of relevant emerging technologies, can continue to demonstrate a multiplier effect by enhancing CE teaching and learning, as well as an accelerator effect for sustainability outcomes at various geographical levels (community, local, national and global). We argue that due to the multiplicity of domains (e.g., engineering, design, economics, business perspectives) and systems thinking involved in CE (Ellen MacArthur Foundation 2018), these emerging interactions and connectedness features for learners, educators, managers, entrepreneurs, and for technologies (e.g., OER-MOOCs interactions, public platform integrations, situating digital learning strategies via TEL strategies) are relevant for developing a CE OOE context and CE MOOC instructional designs. A finding which makes community of inquiry practices around knowledge co-construction relevant for CE education.

4.3. Multi-Level and Multi-Domain Varieties of Agents, Structures and Technologies for a CE OOE Context

Our findings in Tables 3 and 4 also indicate the existence and importance of varieties of structures, internal structures of agents, active agency situations and outcomes that should be taken into account in a CE OOE context and CE MOOC instructional designs. We argue that the structures of existing technological platforms, learning management systems, technological areas, course structures, evaluation structures, societal structures, as well as broader politico-administrative (policies), techno-economic (innovations) and labor market structures (e.g., employability, entrepreneurship) are key determinants for designing competent CE OOE and CE MOOC instructional design solutions. For internal structures of agents, learning relevance, preferences and motivations of learners, as well as for goals of policy makers and administrators as agents aiming at creating/increasing social–ecological returns through education emerge as key dynamics. Active agency situations which involve autonomous study possibility and actualization, active involvement potential, educators' embedding innovations for educating students and citizens, and the goals of educational and CE technology developers and entrepreneurs in commercializing solutions, in addition to their actions on corporate social responsibilities would also be supporting elements for developing a CE OOE context. The outcomes to be targeted range from micro to macro level as awards for participation, sharing resources, developing digital scientific literacy, as well as broader sustainability outcomes, such as developing a circular production and consumption culture among citizens and in overall society. Therefore, we argue that each of these structuration constituents should be taken into serious account in an interacting manner to activate further active agency and improvements in teaching and learning for a CE for the evolution of each type of structure involved in CE OOE and CE MOOC instructional designs.

4.4. Implications for Policy and Financing for CE OOE and CE MOOCs

According to our findings, from policy and financial perspective, OOE and MOOC programs do not pose a threat to the sustainability of other forms of transnational higher education (Wilkins and Juusola 2018). Yet, economic and social sustainability of digital transformation of education become day by day more reliant on the provisions of self-directed, motivating, applicable, rich, and technology-supported designs and implementations on offer (Jin-Hwa and Kim 2016). In this sense, from a market perspective, Porter (2015) argues that a freemium and possible new business models for MOOCs to inform decision-making by managers at universities can be relevant. While the main concerns in this cluster are around scalability and robustness of these business models (e.g., OOE, Business models for Sustainability (Täuscher and Abdelkafi 2018)), the main argument is that MOOCs as a marketing platform remain promising (Tobias Martinez et al. 2016) as they raise, e.g., the profile of the universities (e.g., Northampton Business School, Gateway MOOC) (Anderson et al. 2014). It is also argued that while MOOCs act as part of the information marketing strategy for the universities, they enable the dissemination of various knowledge discourses (e.g., OCL theory) and thematic contents to society (Gallagher 2018) (e.g., sustainability, SDGs, circular economy). In this financial view, benefitting from outsourcing effects, integration and/or interoperability with external virtual learning communities through social networks also opens up a possibility space for cost sharing and cost reductions (Martinez-Nuñez et al. 2016). To reduce costs at micro level, the removal of tutor nodes becomes enabled by different modes of learning driven by participants and within MOOC communities (Mishra et al. 2017). However, we argue that sponsoring, funding or co-funding of OOE for CE teaching and learning can be shared among partners (e.g., municipalities, private companies, non-governmental organizations, universities, both public and private) which target creating social returns (e.g., societal welfare, corporate social responsibility and informing responsible customers, social impact, educating citizens, respectively) on these investments.

For instance, for governments and the education system, private sector, NGOs, even from early childhood, elementary education (e.g., in the scope of corporate social responsibility for private firms and social impact for NGOs) are relevant. In this sense, we agree with Davis (2009) that *"early investments in human capital offer substantial returns to individuals and communities and have a far-reaching effect"*. For instance, Spacebuzz.earth, an early-childhood education organization using virtual and augmented reality technologies to give kids (age range 9–12) the experience of an astronaut via creating an overview effect, and make them ambassadors of our planet, aims at reaching 100 million children to get to the experience of this overview effect yearly in a few years. Or, for instance, in CE context, "Alrededor de Iberoamerica", an educational project in partnership Veolia and Organization of Ibero-American States (OEI) for CE education reached 30,000 students aging between 10 and 11 (Living Circular 2015). In the domain of CE, such scales are very relevant for utilizing a CE OOE and/or CE MOOCs.

For professional education, while CE OOE necessitates incorporating the use of real-world cases via, for example, the contextualization of serious games (SGs), technical aspects of playing these SGs, and debriefing after these SGs. Online versions of such SGs are also shown to help learners shift their personal and professional mindsets, paradigms and practices, such that these shifts are needed for reaching sustainability actions and outcomes that are expected from (and the rationale behind) an education for sustainability (Dieleman and Huisingh 2006). We argue that these mechanisms of creating authentic, contextualized learning tasks, SGs in CE MOOC instructional designs can help CE experts receive additional feedback over (unknown initial state, process, and target states of) complex CE systems through experiential learning (Dieleman and Huisingh 2006). While such applications of experiential learning provide opportunities for encouraging and increasing the participation of various professional agents from various sectors (e.g., business, social, and environmental entrepreneurs) and society at large (e.g., citizens) in learning, the education models of most OOE and MOOCs in the field of education for sustainability we analyzed in this article often lack these possibilities in an integrated manner to deliver a competent and comprehensive OOE design and/or MOOC instructional designs, also for a CE education.

Finally, we argue that to allow for moving forward with developing more open, adaptive, reflective interactions and/or partnership models in education which enable participation of local agents and stakeholders, and which accommodate and recognize their specific needs to create multiple value; OOE, OERs and MOOCs for CE education become key means, and this is in accordance with the discourse and realization of a knowledge society, including an attention to lifelong learning in a context, and with the notion of academic developers as agents of change and partners in arms of change to transform educational practices at socio-technical regime level (Debowski 2014).

5. Conclusions

One of the first and key conclusions of this article is, despite all potential affordances of OOE, TEL, OER, SGs, MOOCs for CE teaching and learning, their potential has not been used to a significant extent in an integrated manner. Our integrated findings and meta-synthesis contribute to new knowledge in relation to the existing research in OOE (and prospectively, educational technology and content design and development) in the field of digital education for sustainability. In this respect, our theoretical lenses, the use of theories of neo-capital, transition management from a multi-level perspective and structuration theory can provide valuable insights for OOE, OCL and CE researchers and scholars, as well as practitioners in these fields, as shown in this article in detail.

Our contribution for development and implementation of a CE culture via research on educational innovations, and via applied educational innovations, firstly comes with the 22 revealed human, social, cultural and technological neo-capital needs for minimum viable CE MOOC instructional designs in the context of CE OOE. Secondly, we further relate these CE MOOC instructional design constituents to 11 technological advances, which can be incorporated in CE OOE to capture and better manage the complexity of CE teaching and learning. Thirdly, we provide broader and evolving multi-level and multi-domain structuration constituents and dynamics among various types of agents, types of structures, and types of technologies, which should be taken into account when designing CE OOE and CE MOOCs to capture micro-level learning preferences and journeys of individuals and groups within broader and evolving meso and macro level technological, organizational and institutional structures. Finally, by doing so, our contribution also informs policymakers' in their potential efforts directed towards combining OOE, TEL, OERs, SGs, MOOCs, technological infrastructures, communities of inquiry (CoI), practice (CoP), and open schooling to foster inquiry skills and lifelong learning for CE (see EC (2019)) in the context of supranational landscape influences. The 46 revealed structuration constituents and, broader and evolving technological, organizational and institutional structures, can help support informed decisions to be taken by national government authorities, specialized agencies in the context of policy text production targeting OOE in a context, and CE education. In the context of practices, academics, researchers, analysts, educators, educational technology developers, entrepreneurs, and innovators can (co-)benefit from our findings in designing and implementing CE OOE and CE MOOCs, as well as in developing and implementing blended learning, customized webcasting facilities, connected apps, linked mobile devices via Internet of Things (IoT) for seamless learning for the context of CE education.

Thus, we recommend that the supply of CE OOE and/or CE MOOC instructional designs, if our integrated findings (22 findings for capital- and neo-capital-based multiple value formations, 11 for emerging tools and technologies, five for micro/meso-level interactions among actors and structures, and nine for macro-level interactions among a broader set of varieties of actors, structures and technologies) are taken into account in developing and implementing such CE OOE and/or CE MOOCs instructional designs, this can further activate a demand-driven process of social and networked learning as a growing niche, and as a gradual transition from teacher-centered to distributed ways of CE learning in which learners are free to learn at any time and any place. These 46 features and selected utilization of their subsets in CE OOE and CE MOOC instructional designs can provide several combinatorial possibilities to educators and learners to interact (with content, with each other, with technologies) and also to learn with and from peers, and help policymakers, administrators, and entrepreneurs decide which type of resources, rules and tools to invest in, create and/or to further develop for a better and competent fit for the interdisciplinary, multidisciplinary or transdisciplinary needs of the processes of teaching, learning and doing for a CE by attaining sustainability outcomes which accelerate CE transition at broader socio-technical regime, and global landscape levels.

Considering our findings and limitations, some future research directions include the following: research on the determinants of learners' motivations in OOE settings in the context of CE; assessment and evaluation of various instructional designs of CE MOOCs; impact of instructional designs of CE MOOCs on the completion rate of learners, on the employability and/or self-employability of learners, and on bringing about circular innovations after completing various CE MOOCs; finally, research on the determinants of the scalability of various CE MOOC instructional designs in the context of CE OOE.

Author Contributions: Conceptualization, S.T. and M.S.; methodology, S.T. and M.S.; software, S.T. and M.S.; validation, S.T. and M.S.; formal analysis, S.T. and M.S.; investigation, S.T. and M.S.; resources, S.T. and M.S.; data curation, S.T. and M.S.; writing—original draft preparation, S.T. and M.S; writing—review and editing, S.T. and M.S.; visualization, S.T. and M.S.; supervision, S.T.; project administration, S.T. and M.S.; funding acquisition, S.T. and M.S

Funding: This research was funded by the United Nations University Maastricht Economic and Social Research Institute on Innovation and Technology (UNU-MERIT) grant number 606UU-907 and the Dutch National Initiative for Education Research (NRO)/The Netherlands Organization for Scientific Research (NWO) and the Dutch Ministry Education, Culture and Science grant number 405-15-705 (SOONER/http://sooner.nu).

Conflicts of Interest: The authors declare no conflict of interest. The funders had no role in the design of the study; in the collection, analyses, or interpretation of data; in the writing of the manuscript, or in the decision to publish the results.

References

Anderson, Terry. 2004. Towards a theory of online learning. *Theory and Practice of Online Learning* 2: 109–19.

Anderson, Maggie, Rachel Fitzgerald, and Ross Thompson. 2014. MOOCs-Mass Marketing for a Niche Audience? Available online: http://nectar.northampton.ac.uk/6790/ (accessed on 30 April 2019).

Aksela, Maija, Xiaomeng Wu, and Julia Halonen. 2016. Relevancy of the Massive Open Online Course (MOOC) about Sustainable Energy for Adolescents. *Education Sciences* 6: 40. [CrossRef]

Griselda Argueta-Velazquez, Martha, and Maria Soledad Ramirez-Montoya. 2017. Innovation in the Instructional Design of Massive Open Courses with Gamification and OER to Train in Energy Sustainability. *Education in the Knowledge Society* 18: 75–96. [CrossRef]

Atkins, Salla, Simon Lewin, Helen Smith, Mark Engel, Atle Fretheim, and Jimmy Volmink. 2008. Conducting a meta-ethnography of qualitative literature: Lessons learnt. *BMC Medical Research Methodology* 8: 21. [CrossRef]

Admssent, Maik, Jasmin Godemann, Gerd Michelsen, Matthias Barth, Marco Rieckmann, and Ute Stoltenberg. 2007. Developing key competencies for sustainable development in higher education. *International Journal of Sustainability in Higher Education* 8: 416–30.

Burch, Sarah L., and Sara E. Harris. 2014. A Massive Open Online Course on climate change: The social construction of a global problem using new tools for connectedness. *WILEY Interdisciplinary Reviews-Climate Change* 5: 577–85. [CrossRef]

Bocken, Nancy M. P., Elsa A. Olivetti, Jonathan M. Cullen, José Potting, and Reid Lifset. 2017. Taking the circularity to the next level: A special issue on the circular economy. *Journal of Industrial Ecology* 21: 476–82. [CrossRef]

Blomsma, Fenna, and Geraldine Brennan. 2017. The emergence of circular economy: A new framing around prolonging resource productivity. *Journal of Industrial Ecology* 21: 603–14. [CrossRef]

Bourdieu, Pierre. 1972. Les stratégies matrimoniales dans le système de reproduction. In *Annales. Histoire, Sciences Sociales*. Cambridge: Cambridge University Press, vol. 27, pp. 1105–27.

Cai, Ken, Yingying Jin, Hongwei Yue, and Haoran Huang. 2016. Analysis of the Learning Mode of an Elaborate Resource Sharing Course. *International Journal of Emerging Technologies in Learning* 11: 66–70. [CrossRef]

Cano, Esteban Vazquez. 2015. Technological Challenge for Massive Open Online Courses Sustainability. *Panorama* 9: 51–60.

Carrera, Jeimmy, and Darinka Ramírez-Hernández. 2018. Innovative Education in MOOC for Sustainability: Learnings and Motivations. *Sustainability* 10: 2990. [CrossRef]

Caruso, Giulia, and Stefano Antonio Gattone. 2019. Waste Management Analysis in Developing Countries through Unsupervised Classification of Mixed Data. *Social Sciences* 8: 186. [CrossRef]

Chen, Yang-Hsueh, and Pin-Ju Chen. 2015. MOOC study group: Facilitation strategies, influential factors, and student perceived gains. *Computers & Education* 86: 55–70. [CrossRef]

Chunwijitra, Sila, Chanchai Junlouchai, Sitdhibong Laokok, Pornchai Tummarattananont, Kamthorn Krairaksa, and Chai Wutiwiwatchai. 2016. An Interoperability Framework of Open Educational Resources and Massive Open Online Courses for Sustainable e-Learning Platform. *IEICE Transactions on Information and Systems* E99D: 2140–50. [CrossRef]

Coelho, José, António Teixeira, Paula Nicolau, Sandra Caeiro, and Vitor Rocio. 2015. iMOOC on Climate Change: Evaluation of a Massive Open Online Learning Pilot Experience. *International Review of Research in Open and Distributed Learning* 16: 152–73.

Corvellec, Herve, and Johan Hultman. 2014. Managing the politics of value propositions. *Marketing Theory* 14: 355–75. [CrossRef]

D'Adamo, Idiano. 2018. The profitability of residential photovoltaic systems. A new scheme of subsidies based on the price of CO_2 in a developed PV market. *Social Sciences* 7: 148. [CrossRef]

Davis, Julie. 2009. Revealing the research 'hole' of early childhood education for sustainability: A preliminary survey of the literature. *Environmental Education Research* 15: 227–41. [CrossRef]

Debowski, Shelda. 2014. From agents of change to partners in arms: The emerging academic developer role. *International Journal for Academic Development* 19: 50–56. [CrossRef]

Dieleman, Hans, and Don Huisingh. 2006. Games by which to learn and teach about sustainable development: Exploring the relevance of games and experiential learning for sustainability. *Journal of Cleaner Production* 14: 837–47. [CrossRef]

European Commission. 2013. Opening up Education'-Making the 21th Century Classroom a Reality. Available online: https://ec.europa.eu/digital-single-market/node/67636 (accessed on 30 April 2019).

EC. 2019. Circular Economy Life Long Learning. Available online: https://circulareconomy.europa.eu/platform/en/news-and-events/all-events/circular-economy-competences-making-case-lifelong-learning (accessed on 30 April 2019).

Ellen MacArthur Foundation. 2019. Available online: https://www.ellenmacarthurfoundation.org/resources/learn/courses (accessed on 4 April 2019).

Ellen MacArthur Foundation. 2018. Learning Landscape. Available online: https://indd.adobe.com/view/a76263e6-f75f-4f12-bbdc-920c01f42c6f (accessed on 30 April 2019).

Erwin, Elizabeth J., Mary Jane Brotherson, and Jean Ann Summers. 2011. Understanding qualitative metasynthesis: Issues and opportunities in early childhood intervention research. *Journal of Early Intervention* 33: 186–200. [CrossRef]

Falagas, Matthew E., Eleni I. Pitsouni, George A. Malietzis, and Georgios Pappas. 2008. Comparison of PubMed, Scopus, web of science, and Google scholar: Strengths and weaknesses. *The FASEB Journal* 22: 338–42. [CrossRef] [PubMed]

Finfgeld-Connett, Deborah. 2014. Use of content analysis to conduct knowledge-building and theory-generating qualitative systematic reviews. *Qualitative Research* 14: 341–52. [CrossRef]

Fini, Antonio. 2009. The Technological Dimension of a Massive Open Online Course: The Case of the CCK08 Course Tools. *International Review of Research in Open and Distant Learning* 10. [CrossRef]

Fitzgerald, Rachel, Maggie Anderson, and Ross Thompson. 2015. Adding Value: Open Online Learning and the MBA. *Electronic Journal of E-Learning* 13: 250–59.

Gallagher, Silvia. 2018. Development Education on a Massive Scale: Evaluation and Reflections on a Massive Open Online Course on Sustainable Development. *Policy & Practice-A Development Education Review* 26: 122–40.

Geels, Frank W. 2004. From sectoral systems of innovation to socio-technical systems: Insights about dynamics and change from sociology and institutional theory. *Research Policy* 33: 897–920. [CrossRef]

Geels, Frank W. 2011. The multi-level perspective on sustainability transitions: Responses to seven criticisms. *Environmental Innovation and Societal Transitions* 1: 24–40. [CrossRef]

Geels, Frank W., and Johan Schot. 2007. Typology of sociotechnical transition pathways. *Research Policy* 36: 399–417. [CrossRef]

Geissdoerfer, Martin, Paulo Savaget, Nancy M. P. Bocken, and Erik Jan Hultink. 2017. The Circular Economy—A new sustainability paradigm? *Journal of Cleaner Production* 143: 757–68. [CrossRef]

Geng, Yong, Jia Fu, Joseph Sarkis, and Bing Xue. 2012. Towards a national circular economy indicator system in China: An evaluation and critical analysis. *Journal of Cleaner Production* 23: 216–24. [CrossRef]

Giddens, Anthony. 1984. *The Constitution of Society*. Cambridge: Polity Press, Oxford: Basil Blackwell.

Gil, David, Jose Fernández-Alemán, Juan Trujillo, Ginés García-Mateos, Sergio Luján-Mora, and Ambrosio Toval. 2018. The Effect of Green Software: A Study of Impact Factors on the Correctness of Software. *Sustainability* 10: 3471. [CrossRef]

Grealy, Freda. 2015. Mobile professional learning for the legal profession in Ireland—A student-centred approach. *Law Teacher* 49: 303–22. [CrossRef]

Gunawardena, Charlotte N., Ludmila Ortegano-Layne, Kayleigh Carabajal, Casey Frechette, Ken Lindemann, and Barbara Jennings. 2006. New model, new strategies: Instructional design for building online wisdom communities. *Distance Education* 27: 217–32. [CrossRef]

Harasim, Linda. 2017. *Learning Theory and Online Technologies*. Abingdon-on-Thames: Routledge.

Harzing, Anne-Wil, and Satu Alakangas. 2016. Google Scholar, Scopus and the Web of Science: A longitudinal and cross-disciplinary comparison. *Scientometrics* 106: 787–804. [CrossRef]

Howarth, Jason Paul, Steven D'Alessandro, Lester Johnson, and Lesley White. 2016. Learner motivation for MOOC registration and the role of MOOCs as a university 'taster'. *International Journal of Lifelong Education* 35: 74–85. [CrossRef]

Kaul, Maya, Maija Aksela, and Xiaomeng Wu. 2018. Dynamics of the Community of Inquiry (CoI) within a Massive Open Online Course (MOOC) for In-Service Teachers in Environmental Education. *Education Sciences* 8: 40. [CrossRef]

Kemp, René. 2011. Ten themes for eco-innovation policies in Europe. *Sapiens–Surveys and Perspectives Integrating Environment and Society* 4: 1–21.

Kolb, David A. 2014. *Experiential Learning: Experience as the Source of Learning and Development*. Upper Saddle River: FT Press.

Lin, Nan. 2017. Building a network theory of social capital. In *Social Capital*. Abingdon-on-Thames: Routledge, pp. 3–28.

Living Circular. 2015. The Circular Economy? Child's Play! Available online: https://www.livingcircular.veolia.com/en/eco-citizen/circular-economy-childs-play-0 (accessed on 30 April 2019).

Santamaría Lancho, Miguel, Mauro Hernández, Ángeles Sánchez-Elvira Paniagua, José María Luzón Encabo, and Guillermo de Jorge-Botana. 2018. Using Semantic Technologies for Formative Assessment and Scoring in Large Courses and MOOCs. *Journal of Interactive Media in Education*. [CrossRef]

Jin-Hwa, Lee, and Heyoung Kim. 2016. Implementation of SMART Teaching 3.0: Mobile-Based Self-Directed EFL Teacher Professional Development. *Journal of Asia TEFL* 13: 331–46. [CrossRef]

Leire, Charlotte, Kes McCormick, Jessika Luth Richter, Peter Arnfalk, and Håkan Rodhe. 2016. Online teaching going massive: Input and outcomes. *Journal of Cleaner Production* 123: 230–33. [CrossRef]

Li, Chao, and Hong Zhou. 2018. Enhancing the Efficiency of Massive Online Learning by Integrating Intelligent Analysis into MOOCs with an Application to Education of Sustainability. *Sustainability* 10: 468. [CrossRef]

Markard, Jochen, Rob Raven, and Bernhard Truffer. 2012. Sustainability transitions: An emerging field of research and its prospects. *Research Policy* 41: 955–67. [CrossRef]

Tobias Martinez, Miguel Angel, Juan Antonio Fuentes Esparrell, Maria do Carmo Duarte Freitas, and Andre Luiz Zani. 2016. Massive Open Online Courses—MOOC as a marketing strategy in universities. *ETIC NET-Revista Cientifica Electronica de Educacion y Comunicacion en la Sociedad del Conocimiento* 16: 349–70.

Martinez-Nuñez, Margarita, Oriol Borras-Gene, and Angel Fidalgo-Blanco. 2016. Virtual Learning Communities in Google Plus, Implications, and Sustainability in MOOCs. *Journal of Information Technology Research* 9: 18–36. [CrossRef]

Mannion, Greg, Gert Biesta, Mark Priestley, and Hamish Ross. 2011. The global dimension in education and education for global citizenship: Genealogy and critique. *Globalisation, Societies and Education* 9: 443–56. [CrossRef]

Mayende, Godfrey, Andreas Prinz, Ghislain Maurice Norbert Isabwe, and Paul Birevu Muyinda. 2017. Learning Groups in MOOCs Lessons for Online Learning in Higher Education. *International Journal of Engineering Pedagogy* 7: 109–24. [CrossRef]

McDowall, Will, Yong Geng, Beijia Huang, Eva Barteková, Raimund Bleischwitz, Serdar Türkeli, René Kemp, and Teresa Doménech. 2017. Circular economy policies in China and Europe. *Journal of Industrial Ecology* 21: 651–61. [CrossRef]

Meinert, Edward, Abrar Alturkistani, Josip Car, Alison Carter, Glenn Wells, and David Brindley. 2018. Real-world evidence for postgraduate students and professionals in healthcare: Protocol for the design of a blended massive open online course. *BMJ Open* 8: e025196. [CrossRef]

Mishra, Danish, Steve Cayzer, and Tracey Madden. 2017. Tutors and gatekeepers in sustainability MOOCS. *On the Horizon* 25: 45–59. [CrossRef]

Mochizuki, Yoko, and Zinaida Fadeeva. 2010. Competences for sustainable development and sustainability: Significance and challenges for ESD. *International Journal of Sustainability in Higher Education* 11: 391–403. [CrossRef]

Okada, Alexandra, and Tony Sherborne. 2018. Equipping the Next Generation for Responsible Research and Innovation with Open Educational Resources, Open Courses, Open Communities and Open Schooling: An Impact Case Study in Brazil. *Journal of Interactive Media in Education* 1: 1–15. [CrossRef]

Orlikowski, Wanda J. 2000. Using technology and constituting structures: A practice lens for studying technology in organizations. *Organization Science* 11: 404–28. [CrossRef]

Oyo, Benedict, Billy Mathias Kalema, and John Byabazaire. 2017. MOOCs for in-service teachers: The case of Uganda and lessons for Africa. *Revista Espanola de Pedagogia* 75: 121–41. [CrossRef]

Papadakis, Stamatios. 2016. Creativity and innovation in European education. Ten years eTwinning. Past, present and the future. *International Journal of Technology Enhanced Learning* 8: 279–96. [CrossRef]

Porter, Sarah. 2015. The economics of MOOCs: A sustainable future? *Bottom Line* 28: 52–62. [CrossRef]

Ramirez-Montoya, Maria S. 2018. Open, interdisciplinary and collaborative innovation to train in Energy Sustainability through MOOCs and educational research. *Education in the Knowledge Society* 19: 11–30. [CrossRef]

Scardamalia, Marlene, and Carl Bereiter. 2006. *Knowledge Building: Theory, Pedagogy, and Technology*. New York: Cambridge University Press, pp. 97–118.

Schophuizen, Martine, Karel Kreijns, Slavi Stoyanov, and Marco Kalz. 2018. Eliciting the challenges and opportunities organizations face when delivering open online education: A group-concept mapping study. *The Internet and Higher Education* 36: 1–12. [CrossRef]

Sneddon, Jacqueline, Gavin Barlow, Sally Bradley, Adrian Brink, Sujith J. Chandy, and Dilip Nathwani. 2018. Development and impact of a massive open online course (MOOC) for antimicrobial stewardship. *Journal of Antimicrobial Chemotheraphy* 73: 1091–97. [CrossRef]

Stahel, Walter R. 2019. *The Circular Economy: A User's Guide*. Abingdon: Routledge.

Steiner, Gerald, and Alfred Posch. 2006. Higher education for sustainability by means of transdisciplinary case studies: An innovative approach for solving complex, real-world problems. *Journal of Cleaner Production* 14: 877–90. [CrossRef]

Stones, Rob. 2005. *Structuration theory*. London: Macmillan International Higher Education.

Täuscher, Karl, and Nizar Abdelkafi. 2018. Scalability and robustness of business models for sustainability: A simulation experiment. *Journal of Cleaner Production* 170: 654–64. [CrossRef]

Tennant, Jonathan P., Harry Crane, Tom Crick, Jacinto Davila, Asura Enkhbayar, Johanna Havemann, Bianca Kramer, Ryan Martin, Paola Masuzzo, Andy Nobes, and et al. 2019. Ten hot topics around scholarly publishing. *Publications* 7: 34. [CrossRef]

Thomas, James, and Angela Harden. 2008. Methods for the thematic synthesis of qualitative research in systematic reviews. *BMC Medical Research Methodology* 8: 45. [CrossRef] [PubMed]

Türkeli, Serdar, René Kemp, Beijia Huang, Raimund Bleischwitz, and Will McDowall. 2018. Circular economy scientific knowledge in the European Union and China: A bibliometric, network and survey analysis (2006–2016). *Journal of Cleaner Production* 197: 1244–61. [CrossRef]

UNESCO. 2015. Rethinking Education: Towards a Global Common Good? Available online: http://www.unesco.org/new/fileadmin/MULTIMEDIA/FIELD/Cairo/images/RethinkingEducation.pdf (accessed on 30 April 2019).

United Nations. 2018. Circular Economy for the SDGs: From Concept to Practice General Assembly and ECOSOC Joint Meeting Draft Concept and Programme for the Joint Meeting of the Economic and Financial (Second Committee) of the 73 UN General Assembly and the UN Economic and Social Council. Available online: https://www.un.org/ecosoc/sites/www.un.org.ecosoc/files/files/en/2018doc/Concept%20Note.pdf (accessed on 30 April 2019).

Wals, Arjen E. J., and Bob Jickling. 2002. "Sustainability" in higher education: From doublethink and newspeak to critical thinking and meaningful learning. *International Journal of Sustainability in Higher Education* 3: 221–32. [CrossRef]

Walsh, Denis, and Soo Downe. 2005. Meta-synthesis method for qualitative research: A literature review. *Journal of Advanced Nursing* 50: 204–11. [CrossRef] [PubMed]

Wang, Ling, Gongliang Hu, and Tiehua Zhou. 2018. Semantic Analysis of Learners' Emotional Tendencies on Online MOOC Education. *Sustainability* 10: 1921. [CrossRef]

Wilkins, Stephen, and Katariina Juusola. 2018. The benefits & drawbacks of transnational higher education Myths and realities. *Australian Universities Review* 60: 68–76.

Wise, Alyssa Friend, and Baruch B. Schwarz. 2017. Visions of CSCL: Eight provocations for the future of the field. *International Journal of Computer-Supported Collaborative Learning* 12: 423–67. [CrossRef]

Zhan, Zehui, Patrick Fong, Hu Mei, Xuhua Chang, Ting Liang, and Zicheng Ma. 2015. Sustainability Education in Massive Open Online Courses: A Content Analysis Approach. *Sustainability* 7: 2274–300. [CrossRef]

© 2019 by the authors. Licensee MDPI, Basel, Switzerland. This article is an open access article distributed under the terms and conditions of the Creative Commons Attribution (CC BY) license (http://creativecommons.org/licenses/by/4.0/).

Article

Retailing, Consumers, and Territory: Trends of an Incipient Circular Model

María D. De-Juan-Vigaray [1,*] and Ana I. Espinosa Seguí [2]

[1] Marketing Department, University of Alicante, 03080 Alicante, Spain
[2] Human Geography Department, University of Alicante, 03080 Alicante, Spain; ana.espinosa@ua.es
* Correspondence: mayo@ua.es

Received: 23 August 2019; Accepted: 21 October 2019; Published: 28 October 2019

Abstract: The aim of this theoretical research is to analyze the state of retail distribution nowadays, reviewing the dynamics of action that contribute to the move from a linear to an incipient circular retail model. The framework is based on the Retail Wheel Spins Theory and the Retail Life Cycle (RLC), with an extra review of Bauman's liquid metaphor. We consider two questions. Firstly, are offline retailers ready to disappear as online commerce and digital marketing aggressively break into the retail industry? Secondly, could commercial spaces (in the fifth stage in the evolution of retail and territory) be in the decline stage in the RLC in the near future or can a circular connection take place? Thus, a desk research methodology based on secondary documentary material and sources issued leads to an interpretive analysis that reveals ten trends (e.g., solid retail vs. liquid retail; glocal retail; food sovereignty) and a wide diversity of changes that could involve offline stores recovering territory and entering a circular phase. Our findings suggest that digitalized physical stores are flourishing and our reflections augur changes in pace and the closure of the linear business cycle to recover territory, the city, its local market, and its symbolism, as well as a liquid business steeped in omnichannel formats developing an incipient circular movement. Conclusions indicate that it is possible to perceive a timid change back to territory and retail spaces which, along with phygitalization, will coexist with the digital world.

Keywords: phygitalization; circular commerce; retailing; digitalization; territory; technology; commercial cycle

1. Introduction

The service sector and, above all, retail activities, as part of the total economic activity of any country, are clearly visible for the society and the territory. As consumers, influenced by cultural elements (Nayeem 2012) we are in charge of obtaining goods and services we need for ourselves, for fulfilling self-consumption, or for other consumers who depend on our purchasing decisions, as families or other social groups. That is why variety, quality and above all, nowadays, convenience and geographical proximity to the retail offer are crucial for the quality of the offline retail service (Kim and Kim 2012). However, consumers are increasingly inclined to purchase more goods online, avoiding the queues in retail stores, facilitating the identification of the best stores via search engines that can provide an infinite assortment of goods or the ability to exhaustively track the product during the delivery process (Kim et al. 2013). This phenomenon is happening around the world with different intensity depending on the country (Ashraf et al. 2014).

The guidelines for retail activity in recent decades have been ruled by the search for maximum competitiveness. Consequently, specific morphological areas of urban areas have been used during particular periods of time and subsequently abandoned once they were no longer profitable, convenient, or attractive for both retailers and consumers. In other words, a linear model of urban space occupancy has been used for retail purposes, without thinking of the consequences on urban degradation.

Focused on Western countries, the invasion-succession processes that took place in North American urban areas were not completely imitated in the majority of European cities and towns, which avoided the functional abandonment of urban centers and the systematic substitution of monospecific retail suburban areas devoted to retailing when becoming no longer competitive, the so called greyfields (Bottini 2005).

This is a time of profound change, and there is a gap in the literature about the forecast linearity when describing the future retail models in commercial distribution (the fifth stage). The strongest and most powerful global dynamics are dominating the market, albeit combined with local and regional experiences of bottom-up retail, which, while not being able to equate in volume terms with the dynamics leading the market, are slowly changing the global business activity perceptions of an increasing number of consumers (Dirlik 2018).

In this commercial context, examining current and future retail trends (Deloitte 2019) and based on the classic cyclical retail theories (McNair 1958; Hollander 1960, 1996) and the Retail Life Cycle (RLC) (Davidson et al. 1976), with the extra review of Bauman's liquid metaphor as a framework, the aim of this conceptual paper is to reflect on the direction retail activities are moving towards nowadays and to look for signs of a new focus, considering the dynamics of action which contribute to a change from a linear to a circular retail model.

Our research questions are two. Firstly, are the offline retailers likely to disappear as online commerce and digital marketing aggressively break into the retail industry? Secondly, could the commercial spaces (fifth stage in the evolution of retail and territory) be in the decline stage in the RLC in the near future or can a circular connection take place? In this sense, the actions of symbiotic recovery between the city and offline retailers are experienced, with the emergence of phygitalization, following the digital tsunami that has led to the eruption of the online phenomenon, and are reviewed.

The paper is organized as follows: firstly, the connections between retail and territory are presented, showing the linear stages. Secondly, the retail offer, demand, and territory are reviewed: the new entrepreneurs and the new business models; the new active, committed, and responsible demand and territory. Next, reflecting on this, the traces of change from a linear commercial distribution model to a circular one are presented. Finally, conclusions and limitations are shown.

2. Connections between Retail and Territory: The Linear Stages

Throughout recent decades, Western countries have witnessed the implication of retail activities in the development of urban areas, being a true reflection of the latest urban dynamics (Espinosa 2011; Bae 2017). It is possible to distinguish four major stages running in parallel to the configuration of the new urban areas, and, above all, to the territorial reconstruction after the excesses and selective abandonment of areas promoted by the law of supply and demand and by the insatiable search for maximum competitiveness among urban spaces, especially those dedicated to retail activities.

In the initial stage, retailing had a markedly local character, formed by a dense and atomized network of small shops distributed throughout the city. Spatial proximity favored the direct and close treatment of consumers, enhanced by the active role that retailers played throughout the process of buying and selling their goods or services (Campesino 1999). Evidently, the urban compactness helped to create this network of shops throughout the city, although it was organized in a hierarchical way with the historic center enjoying full functionality and vitality (Schiller 1994). The progressive shift of retail from the historic center towards the hyper-central areas (Mérenne-Schoumaker 1996) and new suburban centers was forced by various urban, economic, and commercial factors (Dawson 1988).

In the second stage, the city's legacy suffered a progressive social, economic, and architectural deterioration in addition to a clear functional and physical failure to adapt itself to the new retail requirements. This was characterized by a larger area to display products in stores and a gradual reduction in the functions of retailers in the collection and replacement of goods.

In the third stage, the increase in consumer car ownership, the reconfiguration of new retail areas led by great peripheral or commercial driving forces, the economic lethargy of the historic center

(De la Calle 2000), and the relentless aging of the most traditional shops created a territorial map of retailing in which the large national and international companies bought up the majority of the commercial space in hyper-central urban areas, while local and regional shops slowly left their golden age phase. However, as De Mooij and Hofstede (2002) suggest converging technology and disappearing income differences across countries will not lead to homogenization of consumer behavior. Rather, consumer behavior can become even more heterogeneous because of cultural differences.

In this sense, the abandonment of retail activities in historic centers and peripheral districts and the almost exclusive occupation of commercial space by large companies in both suburban and hyper-central areas of cities (Mérenne-Schoumaker 1996; Espinosa 2004) coincided in time.

Despite these forceful, up-down dynamics, it is possible to detect a fourth stage, in which a cycle based on this purchase linear commercial growth is broken to restore the city and the human scale of the act between consumers and their local shops (Vural-Arslan et al. 2011), which is so vital in order to establish long-lasting, sustainable, ethical, and socially responsible relationships (Underhill 2009).

To contextualize this retail cycle, the final phase has a very uneven start. Firstly, from a temporary perspective, there is a large gap between the first actions carried out in North America nearly fifty years ago and the emergence of public and private partnerships in Europe in the past decades (Pattberg et al. 2012). Secondly, Latin America, Asia, and the Arab states for their part, reflect phases that can be placed at different stages (Dávila 2016; Beiró et al. 2018). In North America, urban recovery developed in parallel to suburban retail sprawl and was less rapid and sizeable than in previous times, whereas Western European countries responded more quickly to urban deterioration, due to the value of identity and symbolism of the city's legacy (Turok and Mykhnenko 2007).

That is to say, after decades of suburban growth, but not looking at the degradation and adaptation of urban centers, especially in historic centers, many Western cities tried to bring back vitality to the most depressed historic commercial axes (Rovira et al. 2012). This recovery in the vitality and viability of European urban centers (Schiller 1994; Davoudi 2003) entailed the approach of compactness, which is committed to the development of local retail and the human and experiential scale of the act of purchase. Although there are urban commercial approaches which have decided, in their fourth stage, to wholly commit to this retail urban goal, there are many other examples in which the retail revitalization of central spaces was attempted but balanced with a desire to not limit suburban commercial growth (Ozuduru et al. 2014; Guimarães 2017).

3. Retail Offer, Demand, and Territory

Changes in the retail scene are manifested mainly from the supply and demand sides with substantial differences in retail entrepreneurship and among consumers, such as the impact of consumer green behavior (Dabija 2018) or in other sectors such as luxury retail (Park et al. 2010). In general, it can be said this evolution of the retail distribution sector has been motivated by the interrelated transformation of three indispensable elements: the commercial sector itself, consumers, and the territory. In fact, it is sometimes extremely difficult to know which of these three elements has enabled the paradigm shifts which have taken place in the retail distribution sector.

3.1. The Offer: New Entrepreneurs and New Business Models

The retail offer has experienced a decrease in the number of sales units and retail floor space (Goodman and Remaud 2015). As the consumption of goods and services in households increased, points of sale multiplied throughout the territory in order to better serve consumers, but now the opposite phenomenon is visible. This increase in retail floor space had much to do with the tertiarization of the economies of Western countries and the move from buying to "provision" to just "going shopping" (Roy Dholakia 1999) as a playful and hedonic activity linked, above all, to urban environments (Bell 1976; Shields 1992; Wynne et al. 1998). It was precisely in Western countries that retail expansion had its greatest floor space and economic growth. Moreover, the major companies in commercial distribution operating on a global scale mostly came from countries where they were clearly and successfully positioning themselves for retail in economic tertiarization.

The general trends within the retail distribution sector have been clear (Kaczmarek 2009): the internationalization of supply, the concentration of capital in large distribution chains (mostly of European and North American origin, and starting to be Asian, (Poole et al. 2002)), a decline in independent retailers' market share (Coca-Stefaniak et al. 2005) and the greater presence of franchises and distribution chains in the commercial landscape (Baena 2015; Doherty 2009; Kärrholm et al. 2011; Chkanikova and Mont 2015), a combination of offline and online establishments, the creation of unique experiences in establishments, and the reinvention of the retail outlet through the latest technologies (Deloitte 2019).

Also, over the past two decades, increasingly fierce competition in the retail distribution sector has led to a multitude of acquisitions of small chains by larger distribution groups, causing a reduction in intervention agents in distribution and the disappearance of the weakest companies (Fuchs et al. 2009; Linneman and Moy 2002).

In such an important sector as food, the concentration of the main distribution groups' sales shows the market share of formats such as supermarkets and discount stores (Deloitte 2019). The food sector deserves special attention because of its very frequent consumption and its necessary proximity, or geographical convenience, of the outlets with respect to the consumers (Lavorata and Sparks 2018). Thus, this sector has been made up of an oligopoly of the large food distribution groups controlling a large part of the market share, mainly of packaged food, where there has been a very significant decline of traditional retail outlets since the mid-1990s in all Western countries (Dewitte et al. 2018). The most direct consequence was the expansion of the *"food desert"* (Whelan et al. 2002) territorial problem and the increase in the turnover of medium-sized urban formats, especially supermarkets and discount stores (Wortmann 2004). The outstanding general trend that the retail sector is heading towards is e-commerce, which has dramatically transformed the traditional distribution channel. However, it is also possible to talk about new alternative commercial channels, such as direct selling from producers and farmers, concept stores and consumer cooperatives, in which the consumers themselves design their own offer (Sánchez-Hernández 2009), reinventing the commercial formats of retailing and e-tailing (Turban et al. 2017).

3.2. The Demand: Active, Committed and Responsible Consumers

Demand is the reason for the existence of retailing. Until three decades ago, consumers were perceived in Retail Geography research and Marketing (Spork 1987) as passive agents, although they had the choice of points of sale, taking into account various factors such as proximity to home, value for money, the assortment of goods, the ambience of the store, or the degree of interaction with the owner or sales assistant, being their guides in the act of purchase. In other words, the role of the consumer was limited to selecting the goods available in the stores they relied on for shopping. The explained shopping dynamic is well known, as it prevailed for decades. From the beginning of consumerism, a relationship of dependence evolved between supply and demand, in which the retail offer had to fulfil not just the basic needs of consumers (Baudrillard 2001) but also, create new needs and desires for seducing consumers (Crawford 2004). Moreover, it worked. Over the last few decades, the number of homo consumericus, as mentioned in Saad (2007) in Western societies has dramatically increased.

It is true that the growth in the frequency of purchase and the volume of goods consumed has been an important part of the market machinery, which required an ever faster output of manufactured goods for a faster and more ephemeral consumption (Bell 1976; Gottdiener 1991; Wynne et al. 1998; Farrell 2014). This was aided by the gradual acquisition of cultural capital by a larger number of consumers (Zukin 2004), who carefully studied fashion catalogues and new trends and used the Internet for finding information based on opinion forums and consumer reviews. Penz and Hogg (2011) investigate the multi-dimensional antecedents of approach-avoidance conflicts experienced. They compared online and offline consumers, testing the influence of the situation, product, and reference group on shoppers' intentions shoppers in changing retail environments when purchasing in a particular shopping channel. In general, the crisis was a redefinition of consumer's main features. In short, demand is undergoing a very significant change, meaning that there are more active, committed and responsible consumers who are not malleable by the global dynamics of the distribution sector. A number of consumers have also become sellers via platforms (C2C) and circular upcycling options. What they show by their consumption patterns is that they do not want to consume more, but better, faster, and more conveniently, and, moreover, more healthily.

3.3. The Territory: Location, Location, Location

One of the most universal recommendations shopkeepers receive before opening a retail store is that location is crucial to the success of the business. In fact, Salvaneschi (1996), in his book '*Location, Location, Location*', reaffirms the importance of the territory in retail success. The modernization of retail distribution has had a strong spatial component, since the change from one retail paradigm to another implies the colonization of a new territory without a previous shopping vocation. Likewise, the urban-commercial recovery of the historic centers and central areas of the city has virtually been based on the importance and symbolism that the city has for the city and its inhabitants, the heritage city. As Bottini (2005) stated, when large retail areas were opened on the peripheries, the consumption share was considered, but not the congestion and traffic pollution it caused, nor the emptying of the old centers and offices. Sustainability and environmental aspects were also left aside from urban-retail projects and green marketing was not on the agenda (Groening et al. 2018). However, as pointed out by Fuentes (2011) marketing and consumption of green products has significantly grown in recent years. As a nexus between producers and consumers, retailing plays an important role for the distribution of green products.

In Western countries, and after a long period of excessive growth in which the territory has been exploited as an ephemeral vessel whereby to establish economic activities, a moment of reflection on the future awaiting the city's most symbolic spaces has arrived, and how they could be improved to make them more habitable, functional, and experiential (Raco et al. 2018). Thanks to the great structuring power of the territory having commercial distribution, this reflection on the urban model has had commercial activity as its main support. The return of retail outlets to the historic center of the city represents a much more radical change than the mere transfer of commercial activities to the urban center, a consolidated city district, or to the historic center. It is a change that is about rethinking the social value of this economic activity and its power for the recovery of degraded spaces or those in crisis. It is of paramount importance to steer retail growth towards a better territorial balance and, above all, to restore the human scale of the acts of purchase and to retail in general (Pauwels and Neslin 2015).

4. Traces of Change to a Circular Retail Distribution Model

Countries must identify the stage of their retail landscape to avoid the setbacks that have already been experienced by other territories. The retail industry is showing multifaceted transformations that can be traced to the evolution of space and place (Borghini et al. 2009). When not reacting in time, adjustments to the retail scene and consumption habits will occur. In this sense, there are many initiatives trying to foster local or traditional retail stores as engines of economic and cultural growth and avoid historic urban wastelands. Figure 1 shows the proposed circular commercial distribution model with the trends that are promoting this movement.

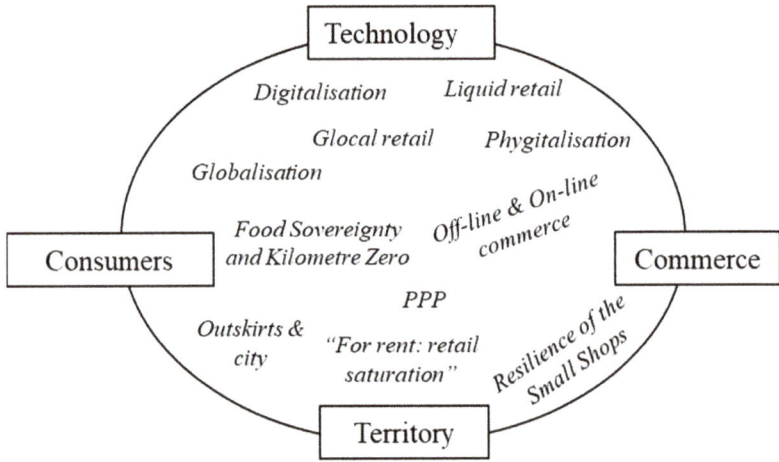

Figure 1. Territory, Commerce, and Consumers Technology circular movement. Source: own devising.

Future of offline retail: Phygitalization. Large retail players operate on an unprecedented global scale because the barriers separating different countries have been significantly reduced. To gain competitive advantage differently in all retail outlets, many innovations are evident from testers in shop windows, payment methods, kiosks, clothing rails that are both real and virtual, interactive touch screens for advertising, to virtual personalized customer experiences. To understand this phenomenon better, researchers have introduced a framework to describe the fusion of the physical and the virtual world in stores: phygitalization. Phygital refers to the concept of both the physical and digital synchronization of the marketing channels to create a different shopping experience (Vel et al. 2015). Lin and Hsieh (2011) explain the concept of phygitalization by bridging the physical store with information technology systems such as tablets and interactive kiosks. The purpose is to create a new-merchandise layout, to make products more accessible, as the customers will be able to order products directly with the digital elements of the store, but also, technology in-store increases the perceived convenience of buying-store, creating interactive and exciting shopping experiences (Blázquez 2014). According to Mitsostegiou et al. (2015), in 2030 retail will be about customer service, virtual stores, interactivity, purchase-home delivery, click & collect, and the traditional distinction between online and offline retail will no longer exist for all age groups and all social groups of the population. The effect of these trends on customer loyalty is beginning to be measured by the large companies. It is no longer a question of providing a service on different platforms, but of combining them. Through the use of technology, there is a fusion of supply, demand and territory.

Solid Retail vs. Liquid Retail. From a solid location to a virtual one: Berman (1983) notes that "*everything solid disappears into the air*". The diffused city evolves towards the compact city. Online retail has inevitably entered into our homes and the solid retail location is virtualized. The point of sale becomes lighter, moves and becomes hybrid. Retail is customized and makes consumers become actors and subjects of its offer, interacting more and more with them, making them participate from start to finish. Retailing is increasingly ephemeral and the point of sale is irretrievably accelerating its life cycle. Shopping is more of a game and is hybridized with leisure, in other words, from 'buying provisions' to simply 'enjoying shopping'. As Herbert et al. (2018) suggest the French food retail industry is experiencing an interregnum. At this point, it is also worth mentioning how virtual retail is looking for a more solid reality, especially in recent years. Warehouses belonging to the most important online retailers are gradually being located in the newly created distribution hubs, near big cities, with similar spatial dynamics to offline retailing.

Digitalization and Globalization in commerce: The changes in grocery stores have taken place with the supremacy of the nearby supermarket against the large hypermarket (Verhetsel 2005; Albecker 2010). This format has become a close geographical area and sensitive to the intra-specific relationship between competition and association, reconfiguring the urban space (Herbert et al. 2018). These places are enabling a reorganization of local stores allowing "chrono-mobility", omnichannel retailing (Zorrilla 2016a), and the optimization of logistics. Sánchez (2012) has shown the changes that the retail sector is undergoing and that we, as consumers, have chosen our mindset when purchasing, and compiling the shopping list has changed. The more traditional stores are trying to reinvent themselves in order to adapt to the new consumer behavior.

One of the biggest risks is in re-inventing something that already exists, just by giving it a tweak of differentiation so that it is not so difficult to be competitive. Cities living in constant change are boosted by two phenomena: digitalization and globalization (Johansson and Öjerbrant 2018). The first has given a boost to many retail companies competing for a better location on the commercial axis of each city (and in SEO positioning). These companies are seeking "agglomeration economies", which account for a massive number of consumers at the respective points of sale, such as shopping centers where an agglomeration of both people and shops can be seen (Rovira 2016). The second one is advancing in leaps and bounds, especially in larger cities. It is true that many medium- and small-sized cities are experiencing local growth and a consolidation of commercial areas, placing them among a large number of cities having commercial areas as poles of attraction (Van Duin et al. 2010). With the assertion of the phenomenon of consumption, the commercial spaces supporting these activities are multiplying. Since it is not possible to imagine a city without consumption, it is equally difficult to imagine consumption disassociated from the shopping center phenomenon.

Retailing on the outskirts vs. retailing in the center and vice versa. Superstores, *"category killers"* or hypermarkets unquestionably chose the outskirts of urban spaces because of their pressing need for space. However, it is observed that these formats are entering urban centers. A well-known German company decided to locate in city centers and an important French category killer specializing in sports followed in its footsteps. Therefore, it seems that there is a trend for smaller and larger space formats to coexist (Reynolds et al. 2007)). The outskirts model has begun to lose its strength and new projects in urban centers are increasingly being proposed. The appeal of the city is its diversity and multifunctionality (Atkinson and Bridge 2004). New, bohemian, attractive and lively residential neighborhoods which are being re-populated in the form of lofts and designer homes incorporated into the run-down areas and neighborhoods (Zukin et al. 2009).

Extending the commercial areas entails additional costs for the cities themselves which means that compact growth is often preferable (Chabrol et al. 2014). From the consumer's point of view, there are various opinions on their preferences. For convenience, many prefer that all the shops are located in a space within everyone's reach without having to make much effort when going shopping (Franzén 2004). This way of thinking makes compact growth a much more attractive proposal, although a totally opposite preference also exists, that is to say, many people, due to the bottleneck caused by all the shops being in a small area leading to a very stressful situation, prefer to travel elsewhere for their purchases or go online (Smith 2002).

'Glocal' (Global and Local) Retail. In the future, "no places" or cloned spaces will be in crisis. Consumers are aware of the fact that the retail offer is being replicated everywhere by the same brands, the same logos, and the same products, leading to a loss of differentiation and authenticity (Foglio and Stanevicius 2007). If consumers close their eyes and are taken to the central square of a shopping center in any city of any country and are asked where they are, the noises, smells, and the atmosphere may very well remind them of any similar shopping center in very disparate locations. If they open their eyes to ask where they are, they will probably be unable to recognize it and definitely confuse it with somewhere they have visited before (Firat 1997).

Today, there is an excessive concentration of chain stores and franchises in all cities, leading to a loss of commercial identity as emblematic and historical shops are disappearing, which irrevocably leads to an extreme similarity between different cities (Chinomona and Sibanda 2013). The competitiveness of small shops in the face of these powerful brands is decreasing and they are now in the background. According to Rovira (2016), it is an intermediate space for local and global retail, that is to say the so-called 'glocal' retail, in which there is an adaptation of the global businesses focused on the reality of local or regional markets. It is an idea put forward by some entrepreneurs who are innovating in the face of changes arising from the shopping habits and lifestyles of consumers.

'For Rent'. Retailing saturation. Today, a quarter of retail premises in Western societies are empty (Saraiva et al. 2019) and an increasing rise in the empty retail surface is likely. It is a problem of saturation which, however, has not been reflected in prices. This will become a serious issue which needs to be managed. The future is committed to highly concentrated commercial spaces. It has become a fact that a proportion of shopping malls is entering a phase of decline in Western countries (Parlette and Cowen 2011).

Firstly in the United States of America, a country where shopping malls have a longer tradition, the so-called "dead malls" are a matter of interest due to their widespread presence (Schatzman 2013). Secondly, dead malls are also being mirrored in Europe and are also facing serious challenges in several countries such as Portugal (Ferreira and Paiva 2017), France (Soumagne et al. 2011), or Sweden (Kärrholm et al. 2011). As for whether there are any surplus stores, effectively there are, but they are missing. What a paradox. The future is committed to a correct adjustment of supply and not an unbalanced growth in it. The existing market is not big enough for the entire supply to be profitable.

The closure of many old premises has attracted a lot of attention and highlighted a reality that has been of concern to many survivors of traditional retail: the large expansion that franchise businesses are experiencing. This phenomenon can be seen in many cities and began to be seen most notably in the 1990s, gaining more strength with the crisis. It is true that nowadays franchises can offer greater stability compared to traditional businesses. That is to say, it is possible to take advantage of the brand's know-how. Many traditional retailers, who were affected during the economic crisis by having to close their shops, and even being dismissed, decided to invest the compensation they received in other safer business sectors (Villaécija 2015).

Off-line Commerce vs. On-line Commerce: E-commerce is an indispensable ally for any company which wants to make its products reach any corner of the world. According to a study undertaken by eBay in 2016 *"The Future of Retail"*, Spain has a global retail growth rate of 27.5%, a not insignificant e-commerce figure (Voces 2016). Retail is determined by a series of political and technological factors which are key to the development of global retail, although both political and economic cooperation between countries has significantly gained weight in order to reduce retail barriers for continuous growth. It is possible to access new markets and remote areas, and often even areas that are somewhat cut off, so that they can consume products which, until now, they could not obtain. With just one click, in addition to being convenient when ordering products which are needed or wanted, costs and time are reduced too, distributors need fewer staff and even less stock, as consumers order the product when they need it and a large warehouse becomes unnecessary (Zorrilla 2016b). Due to our new pace of life and the amenities that online shopping can offer, our shopping habits have changed unexpectedly, in such a way that products are purchased very differently compared to years ago (Gazquez 2016). But what about the territorialization of traditional shops? The fact that online channels exist does not mean we have to neglect traditional retail. Just the opposite, since Amazon has already opened its first bricks and mortar site (Bensinger and Morris 2014). As Fariñas (2016) argues, one of the advantages that make small traditional shops special is the fact that you can touch, smell, and try the product you are thinking of buying. When buying in a physical store the closeness and customer attention received can change the final decision to purchase the product, due to the advice you receive from the staff. On the other hand, the offline retailer adjusts offers and discounts at different outlets to attract the public. Also, they can restructure into more original and specialized formats such as pop-up stores or concept stores.

The Role of Public-Private Partnerships (PPP): PPPs have been established as a mixed management model among private and public agents related to the promotion of central shopping areas, which can have its own legal form that not only affects retailing in a more efficient way but radiates towards virtually the entire public scene where commercial activity takes place (Osborne 2000). Taking as a reference point the stability and efficiency they have demonstrated in the management of urban-commercial space in North America and the countries of Northern and Central Europe, it is possible to reverse commercial decline while beautifying the public space and improving their image for consumers and potential investors (Warnaby et al. 2004; Espinosa and Hernandez 2016).

As The The Economist (2013) argues, many retailers or entrepreneurs take on a variety of tasks ranging from cleaning up areas and their maintenance to security and safety itself. These are complementary services to those offered by town councils or governments on a local scale, thus increasing the perception of cleanliness, safety and the attractiveness of the area. The fact is that when a PPP is implemented, the public administration does not stop providing the public services it has to, but instead delegates some actions which are privatized and run in parallel to public actions. This is the main reason for PPPs being criticized as involving the latent privatization of public management of public spaces (Wieczorek 2004; Vollmer 2008).

More and more citizens are caring about their own cities and the spaces they frequent, thereby recognizing that implementing PPPs can be an important tool to revitalize areas and neighborhoods (Pellicer 2016), potentially attracting many more tourists and visitors, which will make the economy grow and the territory come alive.

The Resilience of Small Shops: Small shops need to implement new strategies to counteract the advance of large stores in cities (Contents 2016; Piqueras Gómez de Albacete 2016). As Fariñas (2016) shows, independent, small shops have many strong points in comparison with large stores: a close relationship with consumers thanks to tradition and character (Doern 2017) and a more flexible capacity of reaction for management and innovation. It should be kept in mind that a number of traditional shops remain "alive" in many urban areas, still having their traditional clientele and attracting new consumers willing to pay for a more individualized customer service (Williams and Vorley 2017). It cannot be denied that in some cases, these consumers are captive citizens unable to move longer distances or to more modern formats due to their own physical limitations or the limitations of the urban areas where they live.

Nevertheless, there is a new type of consumer who needs a homelier shop that does not sell mass-produced products, and, above all, is not pressurized by a shop assistant eager to earn a commission (Fariñas 2016). One of the advantages of traditional shops is the use of local capital which is reinvested at a local scale unlike large chains and franchises, where the money ends up outside these areas (Rosado-Serrano et al. 2018).

Food Sovereignty, zero miles food and local shopping: Sustainability has become a major issue in food production and distribution for a growing number of consumers, who are becoming aware of what is in their shopping basket, the origin of the product, and the environmental and social impacts of food production (Olivares 2017).

That is why a growing number of consumers are demanding their own food sovereignty, which is giving rise to the so-called consumer-led "short marketing channels" or short food supply which have entered into the commercial distribution landscape (Block et al. 2012). Consumers are demanding more local and seasonal products, which have not contributed to the pollution of the environment when they were produced, transported, stored, and commercialized. These include zero food miles products which can be tracked from production to final consumption (Weiler et al. 2014) without going against labor ethics and the environment.

5. Conclusions

Countries and companies, as well as managers and retailers, should consider whether or not they are facing a different commercial paradigm from the one experienced earlier and the effect on the retail wheel while it is still spinning. The commercial scenario is combining phygitalization, glocalization, and omnichannel retailing imbued with territoriality as the new actors of commercial distribution and also of a new space that is physical, but also virtual. Therefore, phygitalization allows digital technologies to act as interactive vehicles whereby consumers can be directly connected to the store itself. Managers must be aware of this, or they will be losing opportunities in the future.

Retail has become liquid. The classic suburban retail formats are pointing at urban centers, reducing their dimensions in order to fit into the central urban retail arena. The resilience of small shops is beginning to bear fruit in large cities where new commercial formats and public and private partnerships in commercial areas (PPPs) are appearing. Sustainability concerns are growing through new distribution formulas and new concepts to include in our shopping list, such as zero food miles and sustainable retail (Atkinson 2013), especially in food distribution, but also related to more ethical concepts such as slow fashion (Naidoo and Gasparatos 2018; Young et al. 2018).

It is expected that by 2025, the business fabric of territorial retail will endure, but will have to be revitalized. At this time, it is possible to say that there will be a crisis for some players and opportunities for others as never before in history. The wheel of retail is continuing to spin (Massad et al. 2011) and future lines of research are open in specific countries to analyze the stage they are at and the factors that will impact on their commercial aspect, trying to avoid earlier mistakes to ensure that commerce, the consumer, and territory co-exist in harmony.

Therefore, the offline retailers are not likely to disappear as online commerce and digital marketing have broken aggressively into the retail industry and commercial spaces—the fifth stage in the evolution of retail and territory—are not in the decline stage in the RLC but a circular connection seems to be taking place.

The limitations of this study are the intrinsic weaknesses of the theories chosen to interpret the traces of the incipient circular model, as well as the lack of empirical data to compare with the stated trends. This new circular model challenges the structure of conventional retailing procedures with bottom-up, local projects run by consumers and small entrepreneurs who defy and bet on a new retail model based on important and necessary sustainable-based values. Thus, environmental awareness, support of local networks of producers and consumers, the importance of retailing with the vitality and viability of symbolic urban areas, and a competitive environment for retailing in which many actors can co-exist are key points of this incipient new way of managing retail activities.

Future research and empirical work can establish if this circular movement is taking place in all countries and continents at the same time and what factors are the key to these trends.

Author Contributions: Conceptualisation: M.D.D.-J.-V. and A.I.E.S.; methodology: M.D.D.-J.-V. and A.I.E.S; investigation: M.D.D.-J.-V.; writing-original draft preparation: M.D.D.-J.-V.; writing-review and editing: M.D.D.-J.-V. and A.I.E.S.; visualization, M.D.D.-J.-V. supervision: M.D.D.-J.-V.; project administration: M.D.D.-J.-V.; funding acquisition (APC): M.D.D.-J.-V.

Funding: This research received no external funding.

Conflicts of Interest: The authors declare no conflict of interest.

References

Albecker, Marie-Fleur. 2010. The effects of globalization in the first suburbs of Paris: From decline to revival? *Berkeley Planning Journal* 23: 102–31. [CrossRef]

Ashraf, Abdul R., Narongsak Thongpapanl, and Seigyoung Auh. 2014. The application of the technology acceptance model under different cultural contexts: The case of online shopping adoption. *Journal of International Marketing* 22: 68–93. [CrossRef]

Atkinson, Lucy. 2013. Smart shoppers? Using QR codes and 'green'smartphone apps to mobilize sustainable consumption in the retail environment. *International Journal of Consumer Studies* 37: 387–93. [CrossRef]

Atkinson, Rowland, and Gavin Bridge, eds. 2004. *Gentrification in a Global Context: The New Urban Colonialism: Gentrification in a Global Perspective (Housing and Society)*. Abingdon: Routledge. [CrossRef]

Bae, Chang-Hee Christine. 2017. *Urban Sprawl in Western Europe and the United States*. Abingdon: Routledge. [CrossRef]

Baena, Verónica. 2015. European franchise expansion into Latin America: Evidence from the Spanish franchise system. *Management Research Review* 38: 149–65. [CrossRef]

Baudrillard, Jean. 2001. *Selected Writtings*. Palo Alto: Stanford University Press.

Beiró, Mariano G., Loreto Bravo, Diego Caro, Ciro Cattuto, Leo Ferres, and Eduardo Graells-Garrido. 2018. Shopping mall attraction and social mixing at a city scale. *EPJ Data Science* 7: 28. [CrossRef]

Bell, Daniel. 1976. *El Advenimiento de la Sociedad Postindustrial*. Madrid: Alianza Editorial.

Bensinger, Greg, and Keiko Morris. 2014. Amazon to open first brick-and-mortar site. The New York City Location to Handle Same-Day-Delivery Inventory, Product Returns. *The Wall Street Journal*. Available online: https://www.wsj.com/articles/amazon-to-open-first-store-1412879124 (accessed on 10 October 2019).

Berman, Marshall. 1983. *All That Is Solid Melts into Air: The Experience of Modernity*. Harmondsworth: Penguin Books.

Blázquez, Marta. 2014. Fashion shopping in multi-channel retail: The role of technology in enhancing the customer experience. *International Journal of Electronic Commerce* 18: 97–116. [CrossRef]

Block, Daniel R., Noel Chávez, Erika Allen, and Dinah Ramirez. 2012. Food sovereignty, urban food access, and food activism: Contemplating the connections through examples from Chicago. *Agriculture and Human Values* 29: 203–15. [CrossRef]

Borghini, Stefania, Nina Diamond, Robert V. Kozinets, Mary Ann McGrath, Albert M. Muñiz Jr., and John F. Sherry Jr. 2009. Why Are Themed Brandstores So Powerful? Retail Brand Ideology at American Girl Place. *Journal of Retailing* 85: 363–75. [CrossRef]

Bottini, Fabrizio. 2005. *I Nuovi Territori del Commercio: Società Locale, Grande Distribuzione Urbanistica*. Florence: Alinea Editorial.

Campesino, Fernández Antonio. 1999. El comercio en los centros históricos de las ciudades españolas. In *Revitalización Funcional del Centro Histórico*. Edited by Begoña Bernal Santa Olalla. Burgos: Universidad de Burgos, pp. 67–83.

Chabrol, Marie, Antoine Fleury, and Mathieu Van Criekingen. 2014. *Commerce et gentrification. Le commerce comme marqueur, vecteur ou frein de la gentrification. Regards croisés à Berlin, Bruxelles et Paris. Le commerce dans tous ses états. Espaces marchands et enjeux de société*. Rennes: Presses Universitaires de Rennes, pp. 277–91.

Chinomona, Richard, and Dennis Sibanda. 2013. When global expansion meets local realities in retailing: Carrefour's glocal strategies in Taiwan. *International Journal of Business and Management* 8: 44–59. [CrossRef]

Chkanikova, Olga, and Oksana Mont. 2015. Corporate supply chain responsibility: Drivers and barriers for sustainable food retailing. *Corporate Social Responsibility and Environmental Management* 22: 65–82. [CrossRef]

Coca-Stefaniak, Andrés, Alan G. Hallsworth, Cathy Parkera, Stephen Bainbridge, and R. Yuste. 2005. Decline in the British small shop independent retail sector: Exploring European parallels. *Journal of Retailing and Consumer Services* 12: 357–71. [CrossRef]

Contents, M. 2016. Tiendas tradicionales: Cómo ganarle a un centro comercial? *Plus Empresarial*. Available online: http://plusempresarial.com/tiendas-tradicionales-como-ganarle-a-un-centro-comercial/ (accessed on 27 May 2019).

Crawford, Michael. 2004. El mundo en un centro comercial. In *Variaciones Sobre un Parque Temático: La Nueva Ciudad Americana y el Fin del Espacio Público*. Edited by Michael Sorky. Barcelona: Ediciones Gustavo Gili SA, pp. 25–46.

Dabija, Dan-Cristian. 2018. Enhancing green loyalty towards apparel retail stores: A cross-generational analysis on an emerging market. *Journal of Open Innovation: Technology, Market, and Complexity* 4: 8. [CrossRef]

Davidson, William R., Albert D. Bates, and Stephen J. Bass. 1976. *The Retail Life Cycle*. Cap. 13. En *Retailing. Critical Concepts*. Edited by Anne M. Findlay and Leigh Sparks. Chicago: Homewood Publishing Company, pp. 264–77.

Dávila, Arlene. 2016. *El Mall: The Spatial and Class Politics of Shopping Malls in Latin America*, 1st ed. Berkeley: California University of California Press.

Davoudi, Simin. 2003. European briefing: Polycentricity in European spatial planning: From an analytical tool to a normative agenda. *European Planning Studies* 11: 979–99. [CrossRef]

Dawson, John. 1988. The changing High Street. *The Geographical Journal* 154: 1–22. [CrossRef]

De la Calle, Vaquero Manuel. 2000. *La Ciudad Histórica Como Destino Turístico*. Barcelona: Editorial Ariel.

De Mooij, Marieke, and Geert Hofstede. 2002. Convergence and divergence in consumer behaviour: Implications for international retailing. *Journal of Retailing* 78: 61–69. [CrossRef]

Deloitte. 2019. Global Powers of Retailing 2019, 22nd ed. Available online: https://www2.deloitte.com/global/en/pages/consumer-business/articles/global-powers-of-retailing.html (accessed on 21 October 2019).

Dewitte, Adam, Sebastian Billows, and Xavier Lecocq. 2018. Turning regulation into business opportunities: A brief history of French food mass retailing (1949–2015). *Business History* 60: 1004–25. [CrossRef]

Dirlik, Arif. 2018. *The Postcolonial Aura: Third World Criticism in the Age of Global Capitalism*. Abingdon: Routledge.

Doern, Rachel. 2017. Strategies for resilience in entrepreneurship: Building resources for small business survival after a crisis. In *Creating Resilient Economies: Entrepreneurship, Growth and Development in Uncertain Times*. Edited by Williams Nick and Vorley Tim. Cheltenham: Edward Elgar Publishing. [CrossRef]

Doherty, Anne Marie. 2009. Market and partner selection processes in international retail franchising. *Journal of Business Research* 62: 528–34. [CrossRef]

Espinosa, Seguí Ana. 2004. Amenazas y nuevas estrategias del comercio de centro urbano. El caso de Alicante. *Boletín de Geógrafos Españoles* 38: 153–74.

Espinosa, Seguí Ana. 2011. La evolución comercial en la ciudad. *Las Actividades Industriales y Comerciales. Implicaciones territoriales*. Available online: http://hdl.handle.net/10045/16301 (accessed on 21 October 2019).

Espinosa, Ana, and Tony Hernandez. 2016. A comparison of public and private partnership models for urban commercial revitalization in Canada and Spain. *The Canadian Geographer/Le Géographe Canadien* 60: 107–22. [CrossRef]

Fariñas, Aránzazu. 2016. Tiendas Online vs. Tiendas Tradicionales. Available online: http://noticias.infocif.es/noticia/tiendas-online-vs-tiendas-tradicionales (accessed on 17 June 2019).

Farrell, James J. 2014. *One Nation under Goods: Malls and the Seductions of American Shopping*. Washington, DC: Smithsonian Institution.

Ferreira, Daniela, and Daniel Paiva. 2017. The death and life of shopping malls: An empirical investigation on the dead malls in Greater Lisbon. *The International Review of Retail, Distribution and Consumer Research* 27: 317–33. [CrossRef]

Firat, A. Fuat. 1997. Educator Insights: Globalization of Fragmentation. A Framework for Understanding Contemporary Global Markets. *Journal of International Marketing* 5: 77–86. [CrossRef]

Foglio, Antonio, and V. Stanevicius. 2007. Scenario of Glocal Marketing and Glocal Marketing as an Answer to the Globalization and Localization: Action on Glocal Market and Marketing Strategy. *Vadyba/Management* 3: 16–17.

Franzén, Markus. 2004. Retailing in the Swedish City: The Move towards the Outskirts. *Insights on Outskirts*. p. 93. Available online: https://orbi.uliege.be/bitstream/2268/62476/1/dynamics.pdf#page=91 (accessed on 21 October 2019).

Fuchs, Doris, Agni Kalfagianni, and Maarten Arentsen. 2009. Retail power, private standards, and sustainability in the global food system. In *Corporate Power in Global Agrifood Governance*. Edited by Clapp Jennifer and Fuchs Doris. Cambridge: MIT Press, pp. 29–59.

Fuentes, Christian. 2011. *Green Retailing-A Socio-Material Analysis*. Lund: Lund University.

Gazquez, Javier. 2016. Las Ventas 'on Line' en España, en la Cuarta Posición Mundial. *El Mundo Newspaper*. Available online: http://www.elmundo.es/economia/2015/02/12/54db4eeeca4741ad788b456c.html (accessed on 13 May 2019).

Goodman, Steve, and Hervé Remaud. 2015. Store choice: How understanding consumer choice of 'where'to shop may assist the small retailer. *Journal of Retailing and Consumer Services* 23: 118–24. [CrossRef]

Gottdiener, Mark. 1991. *Postmodern Semiotics: Material Culture and the Forms of Postmodern Life*. Oxford: Blackwell Publishers.

Groening, Christopher, Josheph Sarkis, and Qingyun Zhu. 2018. Green marketing consumer-level theory review: A compendium of applied theories and further research directions. *Journal of Cleaner Production* 172: 1848–66. [CrossRef]

Guimarães, Pedro Porfírio Coutinho. C. 2017. An evaluation of urban regeneration: The effectiveness of a retail-led project in Lisbon. *Urban Research and Practice* 10: 350–66. [CrossRef]

Herbert, Maud, Isabelle Robert, and Florent Saucède. 2018. Going liquid: French food retail industry experiencing an interregnum. *Consumption Markets & Culture* 21: 445–74. [CrossRef]

Hollander, Stanley C. 1960. The wheel of retailing. *The Journal of Marketing* 24: 37–42. [CrossRef]

Hollander, Stanley. 1996. The wheel of retailing. *Marketing Management* 5: 63–66.
Johansson, Elin, and Frida Öjerbrant. 2018. Mash It Up! Exploring the Phenomenon of Retail Mash-Up and the Survival of the Physical Retail Place in a Digitalized World. Available online: http://lup.lub.lu.se/student-papers/record/8945310 (accessed on 21 October 2019).
Kaczmarek, Tomasz. 2009. Global leaders of retailing and their impact to regional scale. *Geoscape Alternative Approaches to Middle European Geography* 4: 150–66.
Kärrholm, Mattias, Katarina Nylund, and Paulina Prieto de la Fuente. 2011. Retail Resilience in a Swedish Urban Landscape: An Investigation of Three Different Kinds of Retail Places. In *Retail Planning for the Resilient City. Consumption and Urban Regeneration*. Edited by Teresa Barata-Salgueiro and Cachinho Herculano. Lisboa: Centro de Estudos Geográficos, Universidade de Lisboa, pp. 45–62.
Kim, Jae-Eun, and Jien Kim. 2012. Human factors in retail environments: A review. *International Journal of Retail & Distribution Management* 40: 818–41. [CrossRef]
Kim, Jiyoung, Kiseol Yang, and Bu Yong Kim. 2013. Online retailer reputation and consumer response: Examining cross cultural differences. *International Journal of Retail & Distribution Management* 41: 688–705. [CrossRef]
Lavorata, Laure, and Leigh Sparks, eds. 2018. *Food Retailing and Sustainable Development: European Perspectives*. Bradford: Emerald Publishing Limited.
Lin, Jiun-Sheng Chris, and Pei-Ling Hsieh. 2011. Assessing the Self-Service Technology Encounters: Development and Validation of SSTQUAL Scale. *Journal of Retailing* 87: 194–206. [CrossRef]
Linneman, Peter, and Deborah C. Moy. 2002. The Evolution of Retailing in the United States. University of Pennsylvania, Wharton School of Business Working Paper, Volume 443. Available online: http://realestate.wharton.upenn.edu/wp-content/uploads/2017/03/443.pdf (accessed on 21 October 2019).
Massad, Víctor J., Mary Beth Nein, and Joanne M. Tucker. 2011. The Wheel of Retailing revisited: Toward a "Wheel of e-Tailing?". *Journal of Management and Marketing Research* 8: 1–11.
McNair, Malcom P. 1958. Significant Trends and Developments in the Post War Period. In *Competitive Distribution in a Free High-Level Economy and Its Implications for the University*. Edited by Albert B. Smith. Pittsburg: University of Pittsburg Press, pp. 1–25.
Mérenne-Schoumaker, Bernadette. 1996. *La Localization des Services*. Paris: Nathan Université.
Mitsostegiou, Eri, Lydia Brissy, and Alice Marwick. 2015. *Retailing in 2025: What Lies Ahead European Shopping Streets and Malls?* London: Savills World Research Europe.
Naidoo, Merle, and Alexandros Gasparatos. 2018. Corporate Environmental Sustainability in the retail sector: Drivers, strategies and performance measurement. *Journal of Cleaner Production* 203: 125–42. [CrossRef]
Nayeem, Tahmid. 2012. Cultural influences on consumer behavior. *International Journal of Business and Management* 7: 78–91. [CrossRef]
Olivares, Fernando. 2017. Marcas Negras: Cuando el Fabricante no es Quien Pensamos. *El Confidencial. Tribuna*. Available online: http://blogs.elconfidencial.com/economia/tribuna/2017-05-27/consumo-marcas-negras-fabricantes-compra-supermercado_1389600/ (accessed on 15 April 2019).
Osborne, Setephen. 2000. *Public-Private Partnerships: Theory and Practice in International Perspective*. London: Routledge.
Ozuduru, Burcu H., Cigdem Varol, and Ozge Yalciner Ercoskun. 2014. Do shopping centers abate the resilience of shopping streets? The co-existence of both shopping venues in Ankara. Turkey. *Cities* 36: 145–57. [CrossRef]
Park, Jina, Eunju Ko, and Sookhyun Kim. 2010. Consumer behavior in green marketing for luxury brand: A cross-cultural study of US, Japan and Korea. *Journal of Global Academy of Marketing* 20: 319–33. [CrossRef]
Parlette, Vanessa, and Deborah Cowen. 2011. Dead Malls: Suburban Activism, Local Spaces, Global Logistics. *International Journal of Urban and Regional Research* 35: 794–811. [CrossRef]
Pattberg, Philipp, Frank Biermann, Sander Chan, and Ayşem Mert, eds. 2012. *Public-Private Partnerships for Sustainable Development: Emergence, Influence and Legitimacy*. Cheltenham: Edward Elgar Publishing.
Pauwels, Koen, and Scott A. Neslin. 2015. Building with bricks and mortar: The revenue impact of opening physical stores in a multichannel environment. *Journal of Retailing* 91: 182–97. [CrossRef]
Pellicer, Lluís. 2016. El Gobierno Planea Abrir la Gestión de los Barrios a Entidades Privadas. *El País Newspaper*. Available online: http://sociedad.elpais.com/sociedad/2014/06/23/actualidad/1403549621_457228.html (accessed on 16 April 2019).
Penz, Elfriede, and Margaret K. Hogg. 2011. The role of mixed emotions in consumer behavior: Investigating ambivalence in consumers' experiences of approach-avoidance conflicts in online and offline settings. *European Journal of Marketing* 45: 104–32. [CrossRef]

Piqueras Gómez de Albacete, César. 2016. Los Nuevos Comercios y los que Pronto Desaparecerán. November 4. Available online: https://www.cesarpiqueras.com/los-nuevos-comercios-y-los-que-van-a-desaparecer/ (accessed on 21 October 2019).

Poole, Rachel, Graham P. Clarke, and David B. Clarke. 2002. Growth, concentration and regulation in European food retailing. *European Urban and Regional Studies* 9: 167–86. [CrossRef]

Raco, Mike, Daniel Durrant, and Nicola Livingstone. 2018. Slow cities, urban politics and the temporalities of planning: Lessons from London. *Environment and Planning C: Politics and Space* 36: 1176–94. [CrossRef]

Reynolds, Jonathan, Elizabeth Howard, Christine Cuthbertson, and Latchezar Hristov. 2007. Perspectives on retail format innovation: Relating theory and practice. *International Journal of Retail & Distribution Management* 35: 647–60. [CrossRef]

Rosado-Serrano, Alexander, Justin Paul, and Desislava Dikova. 2018. International franchising: A literature review and research agenda. *Journal of Business Research* 85: 238–57. [CrossRef]

Rovira, Agustín. 2016. El Comercio Nuestro de Cada Día: Un Sector Estratégico Que Hace Ciudad. *El Diario*. Available online: http://www.eldiario.es/cv/arguments/comercio-sector-estrategico-hace-ciudad_6_478812122.html (accessed on 15 April 2019).

Rovira, Agustín, D. Forés, and C. Hernández. 2012. *Gestión Innovadora de Centros Comerciales Urbanos. Modelos y Experiencias*. Gijón: Ediciones Trea.

Roy Dholakia, Ruby. 1999. Going shopping: Key determinants of shopping behaviors and motivations. *International Journal of Retail & Distribution Management* 27: 154–65. [CrossRef]

Saad, Gad. 2007. *The Evolutionary Bases of Consumption*. Abingdon: Routledge.

Salvaneschi, Luigi. 1996. *Location, Location, Location: How to Select the Best Site for Your Business*. Edited by Cem Akin. Lembang: Oasis Press/PSI Research.

Sánchez, S. 2012. Nuevos Formatos, Productos Olvidados y Promociones Para Tirar del Comercio. El Mundo Newspaper. Available online: http://www.elmundo.es/elmundo/2012/09/14/economia/1347626084.html (accessed on 15 April 2019).

Sánchez-Hernández, José Luis. 2009. Redes alimentarias alternativas: Concepto, tipología y adecuación a la realidad española. *Boletín de Geógrafos Españoles* 49: 185–207.

Saraiva, Miguel, Teresa Sá Marques, and Paulo Pinho. 2019. Vacant Shops in a Crisis Period. A Morphological Analysis in Portuguese Medium-Sized Cities. *Planning Practice & Research* 34: 255–87. [CrossRef]

Schatzman, Laura. 2013. Metabolizing Obsolescence: Strategies for the Dead Mall. Master's thesis, University of Illinois at Urbana-Champaign, Champaign, IL, USA. Available online: https://core.ac.uk/download/pdf/17355587.pdf (accessed on 21 October 2019).

Schiller, Russell. 1994. Vitality and viability: Challenge to the town center. *International Journal or Retail and Distribution Management* 22: 46–50. [CrossRef]

Shields, Rob. 1992. *Lifestyle Shopping: The Subject of Consumption*. London: Routledge.

Smith, Neil. 2002. New globalism, new urbanism: Gentrification as global urban strategy. *Antipode* 34: 427–50. [CrossRef]

Soumagne, Jean, R. -P. Desse, and A. Grellier. 2011. Commercial Crises and Resilience in French Urban Peripheries. In *Retail Planning for the Resilient City. Consumption and Urban Regeneration*. Edited by Teresa Barata-Salgueiro and Cachinho Cachinho. Lisbon: Centro de Estudos Geográficos, pp. 81–104.

Spork, J. A. 1987. *Vingt-Cinq ans D'étude de Géographie Commerciale au Seminaire de Geógraphie de l'Université*. Actas del Coloquio Internacional Le commerce de detail face aux mutations actuelles. Liège: Les Faits et Leur Analyze.

The Economist. 2013. Bid for Victory. Available online: http://www.economist.com/news/britain/21583704-companies-are-stepping-provide-services-councils-are-cutting-bid-victory (accessed on 17 August 2013).

Turban, Efraim, Jon Outland, David King, Jae Kyu Lee, Ting-Peng Liang, and Deborah C. Turban. 2017. *Electronic Commerce 2018: A Managerial and Social Networks Perspective*. Berlin: Springer.

Turok, Ivan, and Vlad Mykhnenko. 2007. The trajectories of European cities, 1960–2005. *Cities* 24: 165–82. [CrossRef]

Underhill, Paco. 2009. *Why We Buy: The Science of Shopping-Updated and Revized for the Internet, the Global Consumer, and Beyond*. New York: Simon and Schuster Paperbacks.

Van Duin, J. H. Ron, Hans Quak, and Jesús Muñuzuri. 2010. New challenges for urban consolidation centers: A case study in The Hague. *Procedia-Social and Behavioral Sciences* 2: 6177–88. [CrossRef]

Vel, K. Prakash, Collins Agyapong Brobbey, Abdalrhman Salih, and Hafsa Jaheer. 2015. Data, Technology & Social Media: Their Invasive Role in Contemporary Marketing. *Brazilian Journal of Marketing* 14: 421–37. [CrossRef]

Verhetsel, Ann. 2005. Effects of neighborhood characteristics on store performance supermarkets versus hypermarkets. *Journal of Retailing and Consumer Services* 12: 141–50. [CrossRef]

Villaécija, Raquel. 2015. Del Negocio Tradicional a la Invasión de Las 'Ciudades Franquicia'. *El Mundo Newspaper*. Available online: http://www.elmundo.es/economia/2015/08/10/55c77e1f268e3e1f308b4581.html (accessed on 8 October 2019).

Voces, Susana. 2016. El Futuro del Comercio, Según eBay. *Revista Inforetail*. Available online: http://www.revistainforetail.com/noticiadet/el-futuro-del-comercio-segun-ebay/adf8e53a9c90fe19225aed91ff840c0e (accessed on 27 October 2018).

Vollmer, Annette. 2008. Öffentliche und private Interesen in Business Improvement Districts-Zur Frage der demokratischen Einbindung von BIDs in den USA und Deutschland. In *Business Improvement Districts. Ein Neues Governance-Modell aus Perspecktive von Praxis und Stadtforschung*. Edited by Robert Pütz. Berlin: Geographische Handelsforschung, pp. 35–60.

Vural-Arslan, Tülin, Neslihan Dostoğlu, Özlem Köprülü-Bağbanci, and Nilüfer Akıncıtürk. 2011. Sustainable revitalization as a tool for regenerating the attractiveness of an inner-city historic commercial district:'Han District'as a case. *Urban Design International* 16: 188–201. [CrossRef]

Warnaby, Gary, David Bennison, Barry J. Davies, and Howard Hughes. 2004. People and partnerships: Marketing urban retailing. *International Journal of Retail & Distribution Management* 32: 545–56. [CrossRef]

Weiler, Anelyse M., Chris Hergesheimer, Ben Brisbois, Hannah Wittman, Annalee Yassi, and Jerry M. Spiegel. 2014. Food sovereignty, food security and health equity: A meta-narrative mapping exercise. *Health Policy and Planning* 30: 1078–92. [CrossRef]

Whelan, Amanda, Neil Wrigley, Daniel Warm, and Elizabeth Cannings. 2002. Life in a 'Food Desert'. *Urban Studies* 39: 2083–100. [CrossRef]

Wieczorek, Elena. 2004. *Business Improvement Districts. Revitalisierung von Geschäftszentren durch Anwendung des Nordamerikanischen Modells in Deutschland?* Berlin: Technische Uni Berlin, p. 208.

Williams, N., and T. Vorley, eds. 2017. *Creating Resilient Economies: Entrepreneurship, Growth and Development in Uncertain Times*. Cheltenham: Edward Elgar Publishing.

Wortmann, Michael. 2004. Aldi and the German model: Structural change in German grocery retailing and the success of grocery discounters. *Competition & Change* 8: 425–41. [CrossRef]

Wynne, Derek, Justin O'Connor, and Dianne Phillips. 1998. Consumption and the postmodern city. *Urban Studies* 35: 841–64. [CrossRef]

Young, C. William, Sally V. Russell, Cheryl A. Robinson, and Phani Kumar Chintakayala. 2018. Sustainable retailing–influencing consumer behavior on food waste. *Business Strategy and the Environment* 27: 1–15. [CrossRef]

Zorrilla, Pilar. 2016a. El abismo Omnicanal? Para el Pequeño Comercio. Secretos Para Triunfar. Trending Marketing. Available online: https://trendingmarketing.wordpress.com/2015/07/08/el-abismo-omnicanal-para-el-pequeno-comercio-secretos-para-triunfar/ (accessed on 14 April 2019).

Zorrilla, P. 2016b. Ciudades Comercialmente Clonadas o la Muerte de la Diversidad. Trending Marketing. Available online: https://trendingmarketing.wordpress.com/2016/05/22/ciudades-comercialmente-clonadas-o-la-muerte-de-la-diversidad/ (accessed on 14 April 2019).

Zukin, Sharon. 2004. *Point of Purchase: How Shopping Changed American Culture*. New York: Routledge.

Zukin, Sharon, Valerie Trujillo, Peter Frase, Danielle Jackson, Tim Decuber, and Abraham Walker. 2009. New retail capital and neighborhood change: Boutiques and gentrification in New York City. *City & Community* 8: 47–64. Available online: https://eportfolios.macaulay.cuny.edu/benediktsson2013/files/2013/04/Boutiques-and-Gentrification-in-New-York-City.pdf (accessed on 21 October 2019).

© 2019 by the authors. Licensee MDPI, Basel, Switzerland. This article is an open access article distributed under the terms and conditions of the Creative Commons Attribution (CC BY) license (http://creativecommons.org/licenses/by/4.0/).

MDPI
St. Alban-Anlage 66
4052 Basel
Switzerland
Tel. +41 61 683 77 34
Fax +41 61 302 89 18
www.mdpi.com

Social Sciences Editorial Office
E-mail: socsci@mdpi.com
www.mdpi.com/journal/socsci

www.ingramcontent.com/pod-product-compliance
Lightning Source LLC
LaVergne TN
LVHW071953080526
838202LV00064B/6741